Alexandru Calcatinge

The Need for a Cultural Landscape Theory

Stadt- und Raumplanung
Urban and Spatial Planning

Band/Volume 12

LIT

Alexandru Calcatinge

The Need for a Cultural Landscape Theory

An architect's approach

Foreword by HR Prof. h.c. Univ.-Doz. Dr. Peter Jordan

Preface by Prof. Univ. Dr. Arch. Alexandru M. Sandu

LIT

Learning is an endless process.
This book is a small part of it.

Cover images: © by Alexandru Calcatinge

Bibliographic information published by the Deutsche Nationalbibliothek
The Deutsche Nationalbibliothek lists this publication in the Deutsche Nationalbibliografie; detailed bibliographic data are available in the Internet at http://dnb.d-nb.de.

ISBN 978-3-643-90295-5

A catalogue record for this book is available from the British Library

Klosbachstr. 107
CH-8032 Zürich
Tel. +41 (0) 44-251 75 05
Fax +41 (0) 44-251 75 06
e-Mail: zuerich@lit-verlag.ch
http://www.lit-verlag.ch

LIT VERLAG Dr. W. Hopf
Berlin 2012
Fresnostr. 2
D-48159 Münster
Tel. +49 (0) 2 51-620 320
Fax +49 (0) 2 51-23 19 72
e-Mail: lit@lit-verlag.de
http://www.lit-verlag.de

Distribution:
In Germany: LIT Verlag Fresnostr. 2, D-48159 Münster
Tel. +49 (0) 2 51-620 32 22, Fax +49 (0) 2 51-922 60 99, e-mail: vertrieb@lit-verlag.de

In Austria: Medienlogistik Pichler-ÖBZ, e-mail: mlo@medien-logistik.at
In Switzerland: B + M Buch- und Medienvertrieb, e-mail: order@buch-medien.ch
In the UK: Global Book Marketing, e-mail: mo@centralbooks.com
In North America: International Specialized Book Services, e-mail: orders@isbs.com

This book is for my beloved family:

My parents, my grandparents and my twin sister

1 Contents

2 Foreword

This thesis is the result of a comprehensive confrontation with the difficult interdisciplinary concept of cultural landscapes.

From my point of view it is first of all important to note that the relation between man or human community and cultural landscape is a mutual one: on the one hand a certain community with its specific culture shapes and transforms nature and turns a section of geographical space into a cultural landscape - which is in this understanding an imprint of a specific human culture in material and immaterial terms. On the other hand this cultural landscape affects in turn the human community living in it. It has an influence on its attitudes and behaviour, it has a symbolic meaning for it, it conveys traditions and guidelines as much as education or grandmother's tales do. Most people born in or socialised in a certain cultural landscape develop emotional ties to it.

So it is perhaps the task of an architect to conceive and to understand the overall, also the symbolic meaning of a cultural landscape and to place his own works properly in it. This does not mean that the architect must act in full conformity with the inherited cultural landscape and is not allowed to modify it. It could very well also mean to add something different, to set a counterpoint. Counterpoints are sometimes even necessary and required, since they are the expression of new movements and trends in society, also new dominant societal forces.

Nevertheless and under all circumstances, an architect should always be aware of the fact that his work has an impact on the inherited cultural landscape, affirms or modifies it; and that this has in turn an effect on the human community inhabiting this place. So, very simply speaking: he should be aware of the consequences of what he is doing.

On this background, highlighting the concept of the cultural landscape - as this was done by Alexandru Calcatinge - is right in the context of architecture of utmost importance. Every architect should be confronted with this concept and the different views on it. Having laid the foundations for such a confrontation is a great merit of Alexandru Calcatinge's thesis. More specifically I would like to mention the following merits: he has shown that "cultural landscape" is a much disputed concept; he has demonstrated that it is a research topic of various disciplines; he has explained the specific approaches of this disciplines to this concept; he has highlighted the more important views and definitions of this concept; he has emphasized its importance also for achitects, not to speak of urban planners.

Nevertheless, there are also some critical issues with regard to Alexandru Calcatinge's thesis. He uses not a very clear, often metaphoric language which rather mystifies issues. This is, as the philosophical circles he used in his first book, a source to introduce new terms that need to be explained to facilitate their understanding. At some point, the question if this language could communicate this important concept as well as the different views on it to the architects (who are perhaps used to a technical, sober wording) arises. Could this be a step forward towards clarification? I believe that a clarification would come, in this context, as a guide to good practice for architects that could be part of a future research based on the theoretical discussions from this thesis. It is, by all means, judging even by the title, a theoretical research.

Also, as the concept in study is itself controversial and full of contradictions (as Alexandru Calcatinge clearly shows in his thesis), the author's opinions didn't escape contradictions either, but, as F. Scott Fitzgerald once said: "the test of a first-rate intelligence is the ability to hold two opposed ideas in mind at the same time and still retain ability to function"; in this respect, even if the author's language style is metaphorical and its ideas are not free of contradictions, the thesis of Alexandru Calcatinge is a valuable contribution to science insofar as it discusses (very likely for the first time) the concept of the cultural landscape in the context of architecture in a comprehensive way.

Vienna, September 2012.

HR Prof. h.c. Univ. – Doz. Dr. Peter Jordan

Austrian Academy of Sciences
Institute of Urban and Regional Research, Vienna

3 Preface

The doctoral study of the four-year research performed by the architect Alexandru Calcatinge, although it began, as the author says in the introduction, out of an intellectual curiosity, remarked itself in the approach through a evident consequence and stability, from the intention to its finalization. By this very fact is the tenancy of the "long distance" researcher emphasized; this as the research subject would have suggested, at a superficial approach, either a light or explosive commentary (due to well-known or even elementary analysis guidelines) or a syllabic commentary, all the time, in his search for a proper title (repeated changes, due to the many interpretations appeared in the specialized literature, some of them seeming to have nothing in common with the others, offering discussions on very different titles). The consistency of the researcher, whom I followed during his study, did not mean the continuity of an intellectual movement based on a well settled program, as the ideas and under-subjects were arranged and re-arranged, were resumed, were massed and detached by the endless search for a better correlation of ideatic continuation, even if sometimes some commentary relations seem surprising. What in other doctor researches leads through the "study mechanics" to thematic modifications and adaptations, in the case of Alexandru Calcatinge, he fixed the subject of the study giving him importance by its constant approach. This served to sustaining the subject, beyond a certain invasion (even obsolete) of information which risked diluting the object of the main commentary. During my guidance I insisted on the distinction in the approach between the study of the concept of cultural landscape and the study of the cultural landscape, as the difference between the idea and the fact, between understanding and conception, between a philosophy of the implication in the landscape and the method of building the landscape, between situating the gesture in time and space and formalization of the gesture.

Although the interest for a certain cultural landscape or 'generally' for a landscape with cultural values has been found in the specialized literature for a long time (taking as a reference Ortwin Sauer's work of 1925) in the broader use and especially in the space planning, "the cultural landscape" officially enters the practical preoccupation by the documents of the European Union; the same thing happened in fact as in the case of other notional categories, which entered an official use with no clear definition at that moment, being clarified only later in a research about identification and thoroughness towards conceptual definition. To this effect the motivation of the study performed by Alexandru Calcatinge is being sustained in order to respond to the need of having a theory on the concept of cultural landscape.

Although the purpose of the post-graduate refers to the concept of cultural landscape as interfering in the context of the spatial planning, it is evident that understanding this concept is not possible in its complexity without taking into consideration other interfacing interest domains, such as geography, the primary frame, in which the notion of landscape was scientifically forged, at first even under the condition in which the term was first refused by geographers, due to the lack of exactness of its contents, at that moment, which was natural, as on those times the operation with the physical expression and the geometric statement of the expression were specific; and the landscape is obviously a construction of the outer world as well as a visual experience - according to Denis Cosgrove. If associating the image (real and virtual) with the definition of the concept of landscape seems to be essential to me, I do not think that only the social activity can be brought into this definition. On the other hand this activity becomes a condition for the cultural landscape, as a landscape which implies the prints of human existence and experience (so of a social activity) at the level of physical and spiritual visible or registered presence, as former manifestation in the memory of place - condition in which the landscape (cultural landscape) can become an ideological concept. The wealth of the actual interpretations, the tendency towards a more and more pretentious philosophic preciosity of human culture, unjustly leads us sometimes towards the renunciation of simple meanings and their replacement (not to integrate them with newer, more profound meanings). Thus we gave up too easily to taking into consideration the fundamental report (theme of the Congress UIA of 1972 in Varna), we felt it too simple to fugitively explain the report of coverage and differing between landscape and cultural landscape, in order to find ourselves in an ambiguity which the same Cosgrove settled and considered object of confusion regarding understanding and investigating the landscape; it is an ambiguity which we nourish refusing to convene on the distinction between the landscape and the cultural landscape (the first does not have to be a cultural image while the second is!). As we do not stop over the distinction between the object and the perception of the object I doubt that any landscape can be considered “a way to see and perceive the world”, when it does not interfere in the human existence. The human existence is a condition in which the landscape becomes not just a state of the space but a state which reflects a way to see and perceive the world. Under these conditions contemplation is an encouragement towards the accomplishment of the landscape consumption but it is not intrinsic to the landscape quality.

I developed this commentary in order to emphasize the difficulty of analysis (starting from confusion to the preciseness of the research), in which the post graduate got involved with an extraordinary power of investigation, of information and viewpoints gathering out of the specialized literature. Based on this he assumes a difficult task which we present in the introduction as a need of our time: building a theory on cultural landscape. From this work it does not result the satisfaction level of the author about the result. Still, I can appreciate the fact that the work has a considerable consistency of the subject; in the open meaning in the introduction detecting the main aspects of the subject in a thematic approach with no conformism in favour of a relation succession which induce the quantitative and qualitative determinations in history or the actual existence plan, in the factual plan or the conceptual plan, in the conscience plan or the local identity plan - determinations whose convergence towards the ensemble of the cultural landscape is well dominated / explained, even if for some aspects (such as the spirituality expressed only memorially) their understanding stays only suggested.

Although denied by many researchers for whom the fashion of ideas is stronger than the essence of things, the structuralism (as later the phenomenology or situationism) had an important contribution to deepen the understanding of the concept of cultural landscape. Alexandru Calcatinge approaches these determinations, explores them and intelligently exploits them in the scientific plan, with good results in the meaning of some valid conclusions, even if these are left to the reader. As in the text of the work the structural meaning refers for instance to the concept of cultural landscape and the phenomenal understanding to the study of cultural landscape, it lacks of a structurally phenomenological reporting. This gap in the reference of the commentary appears in several sequences of the work serving the ambiguity signaled by Cosgrove and taken over by the post graduate, indeed without damaging the consequence that we observe as being an essential quality of the research and which manages to outrun a certain confusion resulted from this gap - see the paragraph "The methodological study on the concept of cultural landscape" (on the concept of cultural landscape or on the cultural landscape?!). This, although from the point of view of the philosophical understanding of the concept of cultural landscape, is to be remarked in the work of Alexandru Calcatinge a different approach of philosophy as a support for sustaining the concept (although there would have been other philosophical items to be talked about, such as the existentialism or more recent the minimalism) or as an orientation for the initiative to build / develop / promote the cultural landscape. There are two aspects which represent in the work a special relevance regarding the contribution of the author in framing a substantial theory regarding the cultural landscape; this even if in some cases we, as readers, are forced to discern by ourselves when it is about landscape in general and when about cultural landscape. The work of Alexandru Calcatinge develops and completes the personal recommendation of Paul Groth about the descriptive and the interpretative approach, sustaining that accepting the applicability of the landscape studies (cultural landscape!) is constituted politically (meaning public interest) but also personally (with consequences at the level of individual personality).

A first aspect of special relevance for the author's contribution directly implies the constitutive quality of the landscape which results as a consequence of a specific language. Although we have to admit that this specific language is connected not to the landscape in general, but to its quality of being "cultural", that it shows a constitution by description / interpretation and only in the case of the cultural landscape, possibly by birth / construction, the analysis device proposed / commented by the author of the work is the necessary means to knowledge explaining in a systematic relation characteristics of plastic constitutive of space as place, landscape and human existence in the process of landscape definition (and we have to underline that we refer to the cultural landscape!). Lewis' commentary about axioms and about Meinig's principles about those ten possible hypostases of the landscape in the individual's perception help Alexandru Calcatinge to form an exposure motivated on the importance of cultural landscape based on the specialized language, of its expressivity and the valency of this expressivity in the cultural plan. Although it is not very explicitly commentated on (with all the dense incursion, not very necessary regarding the culture concept) from the text it results a well definite opening towards a possible interpretation, even necessary, on the quality of a landscape to be cultural: cultural by what it detects / gathers as marks, cultural by image / addressability, cultural by understanding / evaluation, cultural by getting close / assimilation. Thus the quality of the cultural landscape is a gathering of physical / natural, factual - historical, social-human valences. We should discuss about the particularity of the resulting cultural landscape, about a state of interactivity, defining it beyond of those commented in the text of the work not being evident except of the human existence (dialogues between the individual and the space which Anne Whiston Spirn referred to) and connected to this, existent, outside knowledge.

A second very important aspect for the contribution of the author implies the conditioning through spatialization of the cultural landscape, a conditioning which assumes not the localization of space (as the post graduate says, not very correctly, I say) but the localization in space: we localize an activity, an existence, a configuration, thus a certain part of the space becoming place. We have to emphasize the fact that the cultural valence of a landscape is born once with the localization of an activity in space referring either to an activity with no direct relation to the landscape or an activity implied in the landscape, an activity of interfering with the landscape or just perceiving / contemplating it. Under these conditions we have to say that indeed the architectural space is associated with the process of awareness of space (according to Tuan), but we have to add that is both interactively initial and ulterior, that is a produce of awareness. This information can explain the fact that the meaning of the place (as Jackson observed) is often bound not to the organization of space but to events which take place or which took place in those spaces. It is an observation to another observation which comes to complete with much probability the commentary made by Alexandru Calcatinge regarding the spatialization of the cultural landscape, its determinations and effects / consequences in the human existence. Regarding this aspect it is very interesting the explanation of the role which the cultural landscape takes over through situationalism within the New Urbanism, generating / motivating in space urban and extra urban activities with a character complementary to the work and current existence or with a compulsory characters towards knowledge - both charactering the modern urban existence -

striving to a time situating both of the action itself and the action space.

As the relation space - time is today essential in approaching any process of another state, it is natural that it was not avoided by any analysis of the cultural landscape. Alexandru Calcatinge is attracted by Bakhtin's idea of "chronotope" as a relevant expression for the unit space - time, so that he tries and I think succeeds to convince us that understanding chronotopy in the study of cultural landscape is not just a possibility but a necessity, both conceptually and factually. The post graduate succeeds in dominating the transfer of notions and references form the literary area present in Bakthin's theory towards the physical space of human existence and building, based on the commentaries of several authors, a possible theory of the chronotopy of the cultural landscape in the urban or outside the urban space; a theory which starts from the acceptance of the definition of chronotope as a situation physically defined in space as well as action / happening in a certain time in order to reach the dialogic character of the cultural landscape - actually of the arranged space, and starting from here the presence of chronotopy in the action of human interference over space.

An overlook on the work under the conditions of knowing the further development of the study leads me to the following observation:

The commentaries of the post graduate are mostly of philosophical nature so that failing a reference to what means of classic approach mean, professionally known for the spatial construction, the whole intellectual unfolding (very useful in the space of understanding the essence) will be able to find in a continuous research those necessary guidelines at the level of the real space. Without involving a structural analysis - compositional regarding the composition of the vegetal or built space, the dialogism of the cultural landscape remains an idea, a principle, but it cannot be transferred in the practice of creation and in an environmental existence, as means of expression. There surely are other aspects that with the whole theme, problem, reference wealth and with all the quoted people should have been mentioned in the commentary even if some aspects that are not so important in a normal economy of the text were restrained; I mentioned before the overlapping of the discrepancy between landscape and cultural landscape and I add the lack of distinction clearly enough explained between landscape - concept and landscape - state, between natural landscape and built landscape, as well as re-handling some of the same themes with different commentaries in a perturbed segmentation. The fascination of the wealth of information / documentaries, of the multitude of interpretations or approaches in extending the subject was manifested in the preoccupation of the post-graduate but in a way against selectivity and synthesis. Surprisingly against this themes and documentary wealth the graphic illustration is parsimoniously organized. The photos as well as the schemes (many of which are very consistent as commentary substance) are very convincing, well related to the text, explicit, whether poached, processed or invented by the author.

The critical observations written in the present paper were presented to the post graduate together with my advice in order to make him interfere more personally in tackling some points of view from the specialized literature. I consider that there are serious reasons to take into consideration for neglecting such opportunities especially under the conditions of this new type of doctorate, which dwells on a shorter period for research, which is in this case unadjustable through the enormous effort of the post graduate; let us add the taking up a theme with an exhaustive beach, difficult to dominate in its integrality, the lack of an appetite for approaching little modest things, the temptation of the "philosophical abstract" as an engine of theoretic search and maybe a natural stubbornness, formation or ambition (in research these are needed in a certain way).

Maybe if we had had an easier subject to deal with, such critical observations would not have appeared but Alexandru Calcatinge had and took the chance to involve himself in the research of a difficult subject, somehow controversial, having with all the contributions exposed in the specialized literature a role of a conclusive factor. The post graduate performed to this extend an extraordinary investigation and discovery research, storage and orientation of the information, detecting structural correlations and phenomenological fluxes regarding the definition of the subject; he is proven to tenaciously handle the modern documentation and intelligently use the knowledge means in relation with the interdisciplinary understanding of things. The result: an impressive, responsible research work and an impressive, convincing and useful final work: an important contribution for the theory of architecture and urbanism. All these present Alexandru Calcatinge as a researcher with open chances for innovative approaches.

Bucharest, September 2012.

Prof. Dr. Arch. Alexandru M. Sandu

Professor at the University of Architecture and Urbanism "Ion Mincu"

Former Rector of the University of Architecture and Urbanism "Ion Mincu", Bucharest, Romania (between 1989 – 2000), and former Dean of the Faculty of Urbanism at the same University (between 2000 – 2010).

4 Summary of the study

Introduction

This book is based on my doctoral thesis with the title *Contributions to the theoretical substantiation to the studies of the concept of cultural landscape.* It contains all the major research, except from one chapter that was previously published in my first book. Also, several parts of this book were already published as scientific papers in peer reviewed publications in Romania and Austria by December 2012, or were presented at several international conferences.

Why the need for a cultural landscape theory? Why did I use this title for the book? As it is to be discovered throughout the entire work, the concept is a complicated and ambiguous one. The term cultural landscape is used in different academic disciplines. We use it recently more often in architecture, urbanism and landscape planning. Thus, with the increasing use, we consider it absolutely necessary to know the theory that revolves around the concept. Therefore, this doctoral thesis was born from an intellectual curiosity in terms of knowledge concerning the concept of cultural landscape. The study resulted from a desire to clarify and extend the understanding of the term, which we believe is used inappropriately within our profession. The professional attitude is not totally wrong on this issue, but it often shows a lack of understanding of its theory. In the specialized literature of our country, the concept of cultural landscape is quite rare to be found, and when we found it used in some works, the modality of referring to this concept denotes a minimal involvement of the authors in its study and finding the many meanings this concept has. Thus, as professionals, it is our duty to inform about the evolution and complexity of the concept, being directly involved in its study and development. From the first chapter of the thesis I emphasized the need of a theory of the concept for our profession, in order to prove that we truly master the myriad of different and important approaches in defining and shaping clear what we call *cultural landscape*. So it is important to note that in this research I was tributary to the geographical paradigm of the cultural landscape studies, studying the concept since its first appearance in geography almost one century ago.

The Importance of the Research

I started this research from the premise that good practice is always based on a very well built theory. Thus, the thesis will provide the theoretical basis for a modern approach to the issues of the concept of cultural landscape by facilitating a thorough understanding of concepts, theories and paradigms that led to the development of the concept. Moreover, I propose a series of new approaches, in terms of the new concepts addressed, and this paper will surely be the basis of future research on cultural landscape in our country (or any other country). In this regard, we recognize the importance that architects, planners and landscape architects have in the development of such studies. The proposed thesis will have a profound theoretical character and will analyze the main theories behind the development of the concept of cultural landscape, with personal contributions made to its development by adding new concepts, axioms and theories in the study of the concept.

The Scientific Contribution

While working on the thesis I took into account certain aspects that I considered important in the study of cultural landscape specific to architectural profession. Thus, my contribution to the development of the theory of the concept summarizes to addressing three important elements regarding the theoretical foundations of a concept as complex as that of the cultural landscape. Shortly, I remind that in the first instance I proposed a study of the concept of cultural landscape as a worldview, or Weltanschauung in German. Secondly, I described the role of psycho-geography in the theoretical substantiation of the studies about cultural landscape (this chapter was published in my first book and will not appear in this one). Ultimately I focused my study on the knowledge of the concept of chronotope, analyzing how time and space flow into the theoretical construction of the concept of cultural landscape. The final contribution is to outline the importance of theory in the study of the concept, a number of conclusions being formulated, which are designed to unify these disparate approaches in this paper. So I gave some operational definitions to the concept underlying my approach, and five axioms on the theoretical study of the cultural landscape. All this, together with the partial conclusions that were formulated at the end of the chapters and the short interventions within the construction of the ongoing arguments, are the very basis of my own viewpoint with regard to the need of a cultural landscape theory.

Conclusions

The conclusions of the thesis put into question how the theory helps to understand the concept in all its transformations, relying on the importance of research and the knowledge of the concept as well as understanding its consequences. The conclusions of the rationale of the thesis findings do not want to unify the partial conclusions of the intermediate chapters, but are intended to clarify some aspects encountered during the research and which have not been resolved over time. Thus, the discussion revolves around certain terms encountered during the development of the argumentation, like those of metaphor, palimpsest, innovation, tradition, reinvention, symbol, specificity, sensitivity, perception, interpretation, identity, memory, consciousness and others. I avoid the risk of definitive conclusions, with the aim to leave the confusions, which the concept determined along its evolution to show and conclude on the existence of cultural dynamics underlying the transition and transformability of the concept. I also highlighted the fact that researcher's neutrality is an illusion, and defining concepts is always in relation to other concepts studied in my thesis, and research needs to be related to theoretical reflection. All assumed, according to our findings, an awareness of the changes, highlighting no more and no less than 25 ways of understanding and defining the concept of cultural landscape that we highlighted as the most influential in the academic literature. I believe that there are hundreds of such understandings, that we do not know yet. The final conclusion is based on the formulation of five axioms that I have developed based on the arguments in my thesis as a basis for possible future theoretical studies on the concept of cultural landscape.

Figure 1. The path to knowledge is twisted and blurry, but once you are on top, the whole horizon is opened for you, the researcher, to be found. In the image, view over Mediaş covered in fog in a beautiful September morning. Copyright: Alexandru Calcatinge.

5 Acknowledgments

First and foremost I would like to send my deepest gratitude to my family. I want to thank them for being besides me during all this period of doctoral studies, and for their unconditional support. This thesis is for them, my parents and my sister.

Furthermore, I wish to express my gratitude to prof. dr. arch. Alexandru Sandu, my scientific coordinator for this thesis. Him I must thank for his acceptance to supervise my research during all this time and for his constant constructive criticism and indefinite support for my research. It must be added that I do appreciate the way he managed to focus all my commitments and beliefs for this subject of research on cultural landscapes into one objective and clear goal that is represented by the theory. Although the language we used in constructing the arguments of the thesis combines the narrative and scientific styles, thus being different in comparison with what is used in our profession, I appreciate that professor Alexandru Sandu understood and accepted the fact that those two elements are parts of a young researcher's identity.

In the same way I would like to thank Professor Peter Jordan from the Austrian Academy of Sciences in Vienna, for his invaluable help during the last three years of doctoral studies. He managed to restore my confidence in my own forces. He always found the most appropriate encouragements to determine me to continue a wide research like this one. I must admit that because of his acceptance to be my mentor during my scholarship in Vienna, I was able to experience the richest scientific adventures that life gave me up until now.

Nevertheless, cordial thanks goes to all my colleagues and friends from the University of Architecture and Urbanism „Ion Mincu" in Bucharest, to all the professors and friends from Vienna and from the Faculty of Architecture and Urbanism in Cluj-Napoca. Hoping that this will not offend them, the names of some of them I will acknowledge as follows: prof. dr. arh. Hanna Derer, the one that made everything possible and easy for all the doctoral students; conf. dr. arh. Tiberiu Florescu, for his trust in me for so many years; prof. dr. arh. Constantin Enache, for his warmth and friendship; conf. dr. arh. Cerasella Crăciun, for her trust in me and support for all this time; conf. dr. arh. Cătălin Sârbu, for his trust in me and support, and also for becoming a respected friend; conf. dr. arh. Ioana Mihaela Agachi, for her trust in me ever since my student years in Cluj; Mrs. Argentina Uţă, for her help and understanding for many years. Even if others couldn't find their name here, my respect and gratitude for them I constantly wear.

Last but not least, I would like to thank all the people that helped me translate the thesis from Romanian into English: Ioana Vilcovschi, Nicoleta Cărăbaş, Oana Precup and Paula Bădescu.

6 Introduction

6.1 Argument

Quite recently[1], the concept of cultural landscape has been adopted and used in a considerable number of academic fields, such as architecture and urbanism. Thus, it is necessary that one becomes familiar with the concept before actually using it. However, its use up to the present does not actually imply that it is also understood in all its forms.

This paper is the result of an intellectual curiosity in terms of becoming familiar with the above mentioned concept. One can easily observe a certain tendency to utilize words and expressions that sound good and give refinement to the discourse in the present days. Unfortunately these discourse components are not sufficiently studied and analyzed prior to their use. They are taken as such most of the times, depending on the area they were used before from a single point of view, which reduces their meaning.

The present approach is equally critical to numerous methods of referring to the concept of cultural landscape. The study originated in the dissatisfaction towards how our profession used the concept. As a result, throughout the four years of doctorate we decided to focus on the major importance of the way many theoreticians – geographers, anthropologists, architects, philosophers or urban planners – defined the concept of cultural landscape. And these were all famous theoreticians whose ideas gave birth to ideologies, changed paradigms, and lead to the creation of renowned schools.

The book makes a detailed objective and critical presentation and enquery of a particular part of the theory which gravitates around the operational concept of cultural landscape for the (Romanian) architect, urban planner and landscape architect. But the first fairly asked question regarding the study of this concept would be if we really must develop a theory of this subject more than a century after its first occurrence. Not to mention that in our country the concept was and still is rarely used and debated. And even when we find it mentioned in

[1] This statement demands a short explanation to make it more accurate. One must specify that the statement refers to the study of the concept within our discipline, and no other field such as geography where the concept has long been known. The concept probably acquired a closer study during the past ten years. In this respect, Professor Octav Doicescu tried to facilitate the understanding of the concept from the perspective of the paradigm used in this paper: "the fact that the architect goes beyond his human grouping should not be seen as an exclusively personal or original position. This position can only lead to an isolated creation. It can be the case that such an isolated creation be subsequently adopted by the community out of curiosity for the completion of a cultural landscape. But no creation goes beyond that particular generation or its creator unless it encloses the community's aspirations or expresses the core of perpetual feelings" (Doicescu 1983:25).

certain papers, its analysis is much too poor and the authors prove the minimal involvement in trying to understand the numerous meanings it brings. The role of this process must be made clear because it facilitates comprehension and fundaments a proper use in our approaches. And it is our duty as professionals to provide the correct information with reference to the meaning of this concept. We must in-form the others as well as ourselves since we are directly involved in its study and development. A strong argument in our studies can solely be built on establishing a work hypothesis every time we operate with the concept and also establishing an efficient definition. Some professionals confine themselves to copying existing definitions of the concept established by certain decision forums and therefore do not allow an actual evolution of the study. This is not necessarily a serious problem to face, but a theory on the concept is a more and more important tool for architects, urban planners and landscapers in their practice, on the one side, and for their proper collaboration with geographer, anthropologist and culture theoretician colleagues. All the more so as the growth in this area to an international level is very important and some cultural landscape studies are pluri-, multi- or transdisciplinary. This way we ensure ourselves a well-deserved place among the European and world's most famous professionals.

Under these circumstances do we require a theory on the study on the cultural landscape? If we were to objectively consider the situation within our profession, we have the certainty that such a contribution can only clarify many aspects concerning the way we understand the concept, we analyze and use it within our studies and papers. All this happens in order to demonstrate that we are acquainted with the group of various approaches that are so relevant in defining and outlining the concept. For this purpose, we must specify that our approach is due to the geographical paradigm from the cultural landscape studies. More exactly, we started with the origins of the source, we studied the actual theories that are at the basis of the development of the concept ever since the 20th century. Most of these theories lost their validity and strength in time and had so many completions that they lost the original meaning. We thus encourage the study on cultural landscape which we consider appropriate for our profession. Given the strong connection between architecture and geography, we could constantly notice that the concept of cultural landscape as it came to existence in this field completes the most appropriate form for the studies in architecture and urbanism.

6.2 Scientific role of the study

The subject of the present book was proposed for the Doctor of Philosophy Degree at the University of Architecture and Urbanism in Bucharest. The subject is equally important for the studies in architecture and urbanism in our country as in any other country in the world. The need to know a concept frequently used in the academic and social environments, which was imported from other scientific areas, lead to shaping the doctorate on a number of contributions to the theoretical foundation of the study on the concept of cultural landscape.

Consequently, the paper describes a series of theories from cultural geography, philosophy, urbanism and architecture that concern the concept of cultural landscape. The latter, as demonstrated by our research, was used, developed, polished and brought to the form we currently know it under due to the efforts made by specialists in several different fields. The concept of cultural landscape in urban studies has gained a definite use in specialized literature only after the cultural moment represented by the 1980's and is constantly associated to terms with a strong political character such as globalization, creativity, innovation and regional development. In spite of this, throughout the present paper we will try to determine the main theories that lead to the understanding and change of the term and the concept of cultural landscape based on a multi-level structure. Running through contemporary literature only, i.e. the writings provided during the past ten years, represents a partial phase which can direct the research to a distorted understanding of the term. It is also important that one becomes familiar with the theories from the beginning of the 20th century because this is when the concept actually came to existence.

Therefore, the book will have a strong theoretical character and will present the main theories which helped develop the concept of cultural landscape as well as a series of methods of study of the concept. The purpose is to establish the importance of knowing the theory of the cultural landscape concept as part of its actual study. We propose the professionals in architecture and urbanism a detailed but improvable description of the concept in order to facilitate a complete understanding of a term they are, without any doubt, going to use while exercising their profession.

In this respect, we took into account certain aspects which we considered to be relevant to the study of cultural landscape specific to our profession. Therefore, our contribution to the development of the theory of the concept consists in the approach of three essential aspects that make the basis of such a complex concept as the cultural landscape. Firstly we will suggest a theoretical study of the cultural landscape as Weltanschauung, i.e. briefly, as a worldview. Secondly, the potential role that psichogeography exercises on creating the theoretical basis of the study on the urban cultural landscape was published in my previous book, and will no longer be discussed in this one, even if it was very shortly modified in the final thesis. Thefore, we will focus on the study of the concept of chronotope and the chronotopy process. In this case the analysis will center mostly on how time and space are assimilated in the study on cultural landscape and in the mainly theoretical construction of this concept. All three approaches will have one purpose which is to emphasize the importance of theory in the study of cultural landscape and that as long as the architect and the urbanist are very familiar with the theory, their pluri-, multi- or transdisciplinary studies on cultural landscape are indeed reliable.

The starting point of this research was the idea that a good practice is always based on a well-built theory. For this reason, the present book will set the theoretical basis for a modern approach of the issue of cultural landscape concept by explaining its founding concepts, theories and paradigms. We will also suggest a number of new approaches given the new concepts as mentioned in the previous paragraph. This paper will set up, without any doubt, the foundation of future research on cultural landscape performed in our country and in our profession. In conclusion, we recognize the importance that architects, urban planners and landscape architects in our country have in drawing up such studies.

6.3 Objectives of the study and methods of research

The idea of an approach different than those identified in most of the contemporary specialized papers on cultural landscape occurred in the summer of 2009 when Professor Architect Hanna Derer, Ph.D, took part in the presentation of the first paperwork of the doctorate. With this occasion, she suggested that we should analyze the operative concept of cultural landscape having in view the primary references to the term and thus trying to follow a direction which implies the study of a new theory. Our purpose is to make easily to understand the rich specialized literature and the evolution of the concept. We are not trying to say everything about the cultural landscape nor will this paper be a chronological report on the theories or a historical presentation. We will resort to those theories that help us determine the main aspects of defining the concept, thinking that all analysis of the future theories could easily be misunderstood if the initial references are unknown or not sufficiently studied. At the same time, we will discuss about the connected theories that refer to philosophical concepts which have influenced the paradigms in the field represented by the study on cultural landscape.

The main concern in the present thesis is the presentation of those theories that lead to structuring and defining the concept of cultural landscape, its components and the way we relate to this concept in our profession. We thus propose an original contribution to the development of the theory that gravitates around the concept of cultural landscape and its study. All this in order to complete the generous theory already in use and to propose the architects, the urban planners (the term of urbanologists[2] might actually be more accurate) and the landscap-

[2] The term of urbanologist is not listed in the Romanian explanatory dictionary, but in English the term urbanologist is defined (by the American Heritage Dictionary of the English Language or Webster's New World Dictionary, the 4th edition) as "a sociologist who specializes in the

ers a more extended view of the concept.

With reference to the scientific spirit of a research mentioned by Mihaela Şt. Rădulescu (2011, p.27), the present thesis fulfills at least three essential features. We had in view the following principles as the main (scientific) objectives of the study while drawing out the paper. The first principle refers to "the respect for truth" (Rădulescu, 2011, p.27). Therefore, in our study the respect for truth is translated in its main objective, i.e. to present the concept of cultural landscape in its evolution as well as the importance of the study for the architects, urbanists and landscapers. If we use the term we must also prove that we understand it and this way we show our respect towards the concept which is used probably too often compared to the theory that governs it.

The second principle dwells on "intellectual curiosity" (Rădulescu, 2011, p.27) which is an "inexhaustible source of desire for knowledge manifested as our thought's opening to novelty" (Rădulescu, 2011, p.27). As mentioned previously, curiosity towards the study of the concept was exactly the element that gave birth to this ample research on the concept of cultural landscape, representing a major component in all scientific intervention.

The third principle concerns the "critical spirit" (Rădulescu, 2011, p.28) which "also takes the shape of the need for experimental or theoretical evidence that fundaments rational belief" (Rădulescu, 2011, p.28). This critical spirit can be easily noticed in our approach throughout the entire thesis. Most of the times, we used a critical intervention in our arguments, which fact that can imply a more solid comprehension of the ideas presented in the paper. This can seem difficult for some readers but we consider it important for founding our arguments.

The fourth and last principle is based on "intellectual independence" (Rădulescu, 2011, p.28). This means the use of the well-known authority argument. Thus, Professor Mihaela Rădulescu stated that "a beginner researcher or one who has not yet been recognized for their merits in the scientific world can, of course, resort to authority argument while invoking, in the benefit or against some ideas, the opinions expressed by famous experts" (Rădulescu, 2011, p.29). In this respect, one will observe that in the present paper we will also refer to the contributions brought to the study of cultural landscape by recognized professionals and founders of ideologies who influenced the evolution of the theory around the concept in various disciplines.

Since we mentioned earlier the founding of our arguments in this paper, we must also make a note of the methods of research used. Prior to that though, it is important to remind the readers what Descartes stated about the methods and the four logical precepts that compose it. Therefore, "the first was never to accept anything for true which I did not clearly know to be such; that is to say,

---

problems of cities and urban life". On the www.dictionary.com website one finds the definition of the term urbanology as "the study of urban problems", last visited on 22/08/2012.

carefully to avoid precipitancy and prejudice, and to comprise nothing more in my judgment than what was presented to my mind so clearly and distinctly as to exclude all ground of doubt. The second, to divide each of the difficulties under examination into as many parts as possible, and as might be necessary for its adequate solution. The third, to conduct my thoughts in such order that, by commencing with objects the simplest and easiest to know, I might ascend by little and little, and, as it were, step by step, to the knowledge of the more complex; assigning in thought a certain order even to those objects which in their own nature do not stand in a relation of antecedence and sequence. And the last, in every case to make enumerations so complete, and reviews so general that I might be assured that nothing was omitted" (Descartes, 1999, p.24).

The four precepts analyzed as part of the present paper determine the following aspects. To start with, we propose the analysis of a concept which is unclear and possesses several meanings as well as definitions. This challenged us to focus as much as possible on facts, definitions and coherent explanations of the operative concept in study, with the purpose to facilitate its research and to clarify and simplify its meanings. Although the concept presents itself ambiguous and often superficial in its study, we propose a research which consists in a series of concepts and key terms directing to the development of the theory of the concept of cultural landscape.

Based on Descartes' four percepts, our research is structured mainly on the second and the third percepts. We start with a clear and concise analysis of the terms and concepts (such as cultural landscape, landscape, culture, urban or historic cultural landscape etc.); as we develop our arguments we include in the paper the analysis of new concepts (such as chronotope and chronotopy, dialogism or worldview) and then we draw a more complex analysis referring to understanding, setting the foundations and expanding the theory of the concept of cultural landscape. Finally, our purpose is to determine a favorable frame to the study of the concept, especially on the basis of the theory already developed in the first part of the thesis.

This way, the most complex and generally applied methods of scientific research from our analysis and in founding the arguments are both inductive and deductive (Rădulescu, 2011; Şerbănescu, 2007). Besides these two, the modelisation and analogical models were (more) rarely used because they did not provide exact results in the mainly theoretical study that we undertook. We must specify that the inductive method supposes the structuring of argumentation based on specific information in order to reach a series of general conclusions (Şerbănescu, 2007, p.233). Moreover, the deductive method consists in structuring the argumentation while moving from a general to a particular element which leads to a specific conclusion. The validity of the conclusion is then influenced by the two composing elements of the argumentation, one general and the other particular, also known as major premise and minor premise (Şerbănescu, 2007, p.233). While using the analogical method we made the supposition of the similarities between the concepts, cases and ideas in order to build a precise analogy between them, similar to the inductive method. The integrity of the conclusions results from the similarities between some concepts and not necessarily from their differences. It is the case of a theoretical paper, therefore we have not resorted to this method too often because, as one can easily observe, "the conclusion of the analogical reason, i.e. that from the occurrence of similarities between two objects there must devolve other similarities, brings only verisimilitude, plausibility, but not certainty" (Şerbănescu, 2007, p.233).

At the end of this chapter we would like to make one more note on the style used in the paper and of its features. Probably the most important thing to mention is the use of the so called plural of modesty. Just like Ilie Rad underlined, "one of the conventions of the scientific style is what we call the plural of modesty or the author's plural. (…) However, it is only recommended to students or to those who find themselves at the beginning of a scientific career" (Rad, 2008, p. 41). Also, as this is a translation from the original thesis in Romanian, it might give slightly different meanings to some of the original ideas. Nevertheless, I trust that the main purpose will still remain clear, as I and the ones that helped me translate this thesis did our best to keep the original understanding of the research.

7 Understanding the concept of Cultural Landscape

7.1 The cultural paradigm in defining Landscape

The term of landscape is an extremely complex one. The majority of theoreticians state that it is ambiguous, a statement employed in the branches in which the term is frequently used. However, this very apparent ambiguity and its continuous contextual changing are the ingredients which make it so fascinating from our point of view. The fact that the term is in a permanent and complex evolution might be the very reason for which its ambiguity is accepted as reference.

In the process of time, the term has no longer been used only to express artistic representations or natural frames. Today we use terms such as architectural landscape, urban landscape, political landscape, cultural landscape, media landscape. But not all of these phrases are directly connected to the initial term of landscape. However these phrases are used to name a contextual unit. Thus the term has become a stereotype, a cliché.

In Romanian, the term *peisaj* comes from the French *paysage* (in French, the word pays means country, county, region) and the Latin *pagensis*. The geographer Lucian Drăguţ considers landscape as an important concept which "has a special place among the concepts which connects us to our living environment. (...) The current usage of this term suggests the image of a whole made of dynamic elements, each having its own expression and role in the general context" (Drăguţ, 2000, p. 7). However, the author presents a history of the notion of landscape closely connected to the terms of landscape design or landscape painting, but in the works of Anglo-Saxon origin the approach is slightly different and, in our opinion, easier to assimilate in the understanding of the related term of cultural landscape.

Thus, Denis Cosgrove confirms the fact that the term landscape is ambiguous and imprecise in its geographical usage[3] (Cosgrove, 1998, p. 13). For the author, landscape is not only an image of what exists in nature. It is a construction of the external world by means of social activity and human experience[4] (Cos-

---

[3] The original quote: "In geographical usage landscape is an imprecise and ambiguous concept whose meaning has defied the many attempts to define it with the specificity generally expected of a science. While landscape obviously refers to the surface of the earth, or a part thereof, and thus to the chosen field of geographical enquiry, it incorporates far more than merely the visual and functional arrangement of natural and human phenomena which the discipline can identify, classify, map and analyze".

[4] The original quote: "In other words landscape denotes the external world mediated through subjective human experience in a way that neither region nor area immediately suggest. Land-

grove, 1998, p. 13). Landscape, according to Denis Cosgrove's thesis, is both a social product and a visual experience. It is a social product due to human intervention, and secondly it represents a visual experience due to the artistic part of using the concept of landscape[5] (Cosgrove, 1998, p. 14).

At the same time, an important aspect emphasized by Denis Cosgrove is connected to Richard Hartshorne's intention to remove the term of landscape[6] from its current usage in geography as long as one cannot attribute a coherent scientific definition to it[7] (Cosgrove, 1998, p. 15). One can observe, however, that the term has managed to break the barriers of various fields and domains and it has started to be used more and more in the professional studies. This fact has led to deepen the ambiguity with which it was associated at the beginning. Thus landscape has become an "ideological concept"[8] (Cosgrove, 1998, p. 15).

Another example would be the way in which Peter Longatti and Thomas Dalang propose a linguistic study of the term landscape which appears in the Swiss Government official documents. This is precisely in order to enable the understanding of the term and to remove any inherent ambiguities of using this phrase (Longatti, Dalang, 2007, p.35). Thus, the authors state that merely looking up in the dictionaries to see what the term landscape means is not sufficient. But this step is necessary. Therefore the meaning of landscape is determined by the very context where it occurs. Hence, their hermeneutic approach will determine the meaning of the term landscape in relation to other terms depending on the context where they appear[9] (Longatti, Dalang, 2007, p. 36).

---

scape is not merely the world we see, it is a construction, a composition of that world. Landscape is a way of seeing the world".

[5] The original quote "The frequent association in geographical writing of landscape with studies of the impact of human agency in altering the physical environment serves to remind us that landscape is a social product, the consequence of a collective human transformation of nature. The elision of landscape with wilderness or nature untained by human intervention is a recent idea generally involving a rejection of the evidence of human action. The artistic use of landscape stresses a personal, private, and essentially visual experience".

[6] In this case, and as a consequence of those mentioned above, we must ask ourselves if the initiative might be applied also in the case of the term of cultural landscape. This term has come to be used now in numerous various fields, giving the impression that it eludes any attempt to make it easier to be understood and used. At the same time, however, we have noticed from what we mentioned earlier that the two terms, those of landscape and cultural landscape, are often perceived approximately in the same way.

[7] The original quote: "The unities and tensions of subject and object, and of personal and social, produce problems for those geographers who would wish to derive from landscape a specialized scientific concept. Such an attempt requires by its nature the removal of ambiguity, so that scientific understanding may proceed. So with characteristic logic Richard Hartshorne (1939) argued for landscape's exclusion from the geographical vocabulary unless its meaning was so refined as to expunge all subjective and personal connotations".

[8] The original quote: "Landscape, I shall argue, is an ideological concept. It represents a way in which certain classes of people have signified themselves and their world through their imagined relationship with nature, and through which they have underlined and communicated their own social role and that of others with respect to external nature".

[9] The original quote: "Thus we use the hermeneutic approach which is based on the principle that we cannot understand the parts of a text until we understand the whole and vice versa. This

The authors propose an analysis of a text of the Swiss constitution which makes reference to the preservation of the natural and cultural patrimony. Thus the first version of 1962 is compared to the modified one of 1999 and it is pointed out the way in which the changes operated on the text of the constitution show the various meanings of the term landscape depending on the context where it appeared (Longatti, Dalang, 2007, p. 37). Thus, we can make out two major differences in the last analyzed version. According to the two authors, the meaning of the two phrases that were eliminated or changed in the constitution text also influences the comprehension of the concept of cultural landscape as well as the meaning of what we understand by the phrase sense of place. Thus we will refer to the terms of *Heimat/ heimatlich* and *Landshaftsbild*. The term *heimatlich* means feeling at home, and *Heimat* means home or native land. *Landshaftsbild* means landscape image or scenery. The modifications of these terms have led to a change of context by removing the subjective and interpretative side of the text, and the new terms that are used give another meaning also to the term of landscape. Thus, as the authors state, landscape is no longer considered as an important and basic element of the interhuman relations, but rather as an inheritance that must be protected[10] (Longatti and Dalang, 2007, p. 37).

Figure 2. Lake Voelkermarkt in the South of Austria. Copyright: Alexandru Calcatinge.

means that we need to look at how the word 'landscape' is related to other terms in a particular text. From these relations we derive the meaning or implied meanings of the word. In legal texts such as the ones dealt with in this study we have the additional benefit that the term must actually be defined exactly when it comes to the implementation of the law or regulation. When an inventory had to be drawn up listing all the mire landscapes of national importance and these landscapes had to be delimited on a map, the surveyors had to have a set of rules with which they could decide where to draw the border of each landscape".

[10] The original quote: "landscape is no longer seen as the home of people but as an inherited asset which could be lost and has to be safeguarded".

Landscape has thus become a resource that can be exploited and which can generate gains. In the race of globalization and of technological supremacy, natural resources have become pillars of economical relaunch and of political discourse, warranty for an evolution based on the regeneration of the Planet's resources. At the same time, landscape remains a social construct, and according to Denis Cosgrove, merely an idea (Cosgrove, 1998, pp.13-38). Perhaps this is precisely why it has had lots of simplistic, but also varied, approaches. But, as Denis Cosgrove stated, the ambiguity of this term and concept which is the very core that has to make us continue the process of understanding and investigating landscape[11] (Cosgrove, 1998, p. 15). For Cosgrove, landscape is not merely a way of seeing and perceiving the world, but also a social and cultural product. Therefore, landscape is a mental construct and a cultural image of our living environment, and its understanding implies the process of contemplation, precisely to the purpose of defining it and using is as scientific term.

Figure 3. A model for contemplating the (cultural) landscape. In this image, a panorama over Mărişel village in Cluj county, Romania. Copyright: Alexandru Calcatinge.

However, we mentioned the process of contemplation precisely because of the direct connection we think it has to the process of reading landscape and, consequently, to the way of relating to this concept and to its definition. The state of contemplation is associated to the artistic aspect of comprehending the concept of landscape. Contemplation is directly connected to perception and human experience. Even the mere look at landscape involves both senses and intellect and it is the first level of the contemplating process. This process implies a deliberate action which determines a private and direct communication to the scene which is being contemplated. We witness the individual's transposition from the beholder's role into the actor's role. Thus there appears its direct involvement in the process of identifying and interpreting landscape[12] (Cosgrove, 1998, p. 2). One must also consider the fact that the process of contemplation is a subjective and interpretative one. Its result is different from an individual to another and it is difficult to quantify and measure. The ambiguity of the term and of the concept of landscape rests also in the very way of perceiving it, as we have already seen.

[11] The original quote: "To accept the ambiguity and severally-layered meanings of landscape does not excuse us from careful examination of them and of their origins".
[12] The original quote: "Contemplation indicates a deliberate attention, often implying a concentration on ideas, objects, or places that are somewhat outside our day-to-day thoughts".

As a conclusion regarding landscape identification and interpretation, it is worth mentioning the discussion carried with professor Peter Jordan on this topic of perceiving and contemplating landscape, of creating that sense of place which we will mention again in this work. There is an important matter which laid the basis of our discussion: what determines this sense of place, as far as the individual is concerned? This issue is not extremely difficult to approach if it is meant to be treated at a simplistic level, but it can make the topic of some broad research if we wish to further clarify the theme, but not exhausting it. The way in which the individual reacts as far as landscape is concerned is important in determining a sense and in forming an affiliation towards a place. Professor Peter Jordan's approach was extremely clear, considering as defining elements in this process, among others, the individual's birth place and experience. This fact was also asserted by Yi-Fu Tuan, Denis Cosgrove, J. B. Jackson and others. Therefore I did not hesitate in bringing up in our dialogue the aesthetician Titu Popescu's argument, who presents the very exception confirming the rule, according to which landscape is a social and cultural construct, subject to the interpretation that is specific to each and every individual. Thus Titu Popescu explains his own situation: as he was born in Botoşani and growing up in the region of Alba Iulia and Sibiu, he does not have any attraction towards the natural, but on the contrary, it does not raise any sort of interest in him (Popescu, 2009, p.11). The nature triggers no interest in him (or the natural beauty, as we discover while reading his work), just the same as, in its turn, nature is indifferent to the individual. In our opinion, this indifference is not a given fact, but it is again a mental construct of the individual. And so is the indifference of nature towards the latter.

Hence, our correlation to landscape is based on our own intellect. There are, indeed, persons who remain indifferent to landscape, but there are also persons who, by beholding nature, remain impressed and ecstatic, without having a clear explanation for their reaction. Here is how the process of reading and interpreting landscape is based on perception and experience and these two, in their turn, are based (among others, as we will see later on in this work) also on contemplation. And contemplation is however based on what was called acquaintance in Bertrand Russel's perception, in Romanian "*cunoaştere*"[13] (Miah, 2006, p. 26).

[13] The original quote: "It must be noted here that when Russell calls 'acquaintance' the simplest and most pervading aspect of experience, he uses 'experience' in a pre-theoretical sense (in line with the language of daily life). But soon he starts using it as equivalent to acquaintance. His discussion of how far our experience can reach is exactly the same as his investigation of the nature of objects of acquaintance. Thus he refers to the objects to which at a given moment he could, as he states, 'give proper names', as 'the objects of my 'awareness', the objects 'before my mind', or the objects that are within my present 'experience''".

7.1.1 Some principles in defining the concept of Landscape: a cultural perspective

Knowing the study framework as far as cultural landscape is concerned represents a challenge for most of those involved. In the professional literature there are three important names in the human and cultural Anglo-Saxon geography who managed to clearly mark out the theoretic context of the studies of cultural landscape in the second half of the 20th century: Pierce Lewis, Donald W. Meinig and Paul Groth. They continued the studies of the well-known Carl Ortwin Sauer and J. B. Jackson as well as of Yi-Fu Tuan in the field of cultural landscape.

Along with them we might mention Anne Whiston Spirn, Christopher Tilley, Tim Ingold or Denis Cosgrove and many others who, by their ingenuity and devotion contributed to the comprehension of an operative concept which, even after a century of usage, managed to keep its vitality which consecrated it in the field of human and cultural geography. Thus, we will present in the next part the theories of those mentioned above, theories that laid the bases for the comprehension of the cultural landscape by what we called the cultural paradigm of landscape studies.

Figure 4. The village of Biertan, Sibiu county. A typical saxon village on Târnava Mare valley. Copyright: Alexandru Calcatinge.

Thus, Paul Groth[14], an architecture and geography professor at Berkeley University in California, underlines the fact that the term landscape in the case of cultural landscape studies means not only a natural framework, but also the interaction between man and space, the relation individual – place, the particularities of the spaces in which social groups coexist and in / out of which they form their identity[15] (Groth, 1997, p.1). The professor's interest on the study of the concept distinguishes itself exactly by the way in which we can comprehend the cultural landscape and his indignation towards where the cultural landscape studies are heading now is obvious as early as the beginning of his work Frameworks for Cultural Landscape Study, in which he mentions the cultural ignorance towards landscape that was manifested even in the cultural landscape studies in the United States[16] (Groth, 1997, p. 1). Paul Groth finds important the study that we must approach regarding the actual and everyday landscape, and this approach can be done in two ways: descriptive and interpretative. Both ways require a careful observation of the cultural landscape which would lead to a coherent analysis based on a series of information which, after a correct filtration, could determine a comprehension of it as complex as possible. The duplicity character of the landscape implies both an intelligent description and a theoretic generalization, which, however, would not be confined to a definition of terms leading to a limitation and to a too strict categorization of the operative concept. Thus, as Paul Groth stated, "whether academic or not, the overarching objective of cultural landscape writing is to inform the public, and the potential applications of landscape studies are political as well as personal"[17] (Groth, 1997. p. 10).

7.1.2 The existence of a Landscape language

As it is generally (re)known, the concept of cultural landscape has also presented a certain ambiguity from the very beginning; the repeated attempts to mark

[14] The author uses the term of tenet which means doctrine, dogma, principle, theory, thesis, opinion. Therefore the choice of the term basic principles in the title belongs to us.

[15] The original quote: "For writers in cultural landscape studies, the term landscape means more than a pleasing view of scenery. Landscape denotes the interaction of people and place: a social group and its spaces, particularly the spaces to which the group belongs and from which its members derive some part of their shared identity and meaning. All human intervention with nature can be considered as cultural landscape: the high-style cathedral or office tower, as well as the Depression-era Hooverville hut, a farmer's barbed-wire fence, or a kitchen garden. Cultural landscape studies focus most on the history of how people have used everyday space – buildings, rooms, streets, fields, or yards – to establish their identity, articulate their social relations, and derive cultural meaning".

[16] The original quote: "For almost fifty years, several loosely allied group of writers and scholars have challenged such cultural ignorance in the United States. Most of them have done this under the rubric of cultural landscape studies".

[17] Paul Groth stated in the introduction of the volume Understanding Ordinary Landscapes.

out what is not directly visible have led to the discovery of some major relations and causes in the control and transformation of landscape, emphasizing the fact that landscape is culturally determined.

The processes of reading and decoding cultural landscape aim at finding the hidden processes that formed it and on which the existence of symbols is based. Hence, the comprehension of landscape is determined as a text, a social and cultural document which, as its reading goes on, will uncover the layers of the values and processes accumulated over time, in a set of "reading between the lines". The form of reading cultural landscape is thus related both to its immediate, visual content and to the mental and sensible content of the individual or collective memory.

During the processes of reading and interpreting cultural landscape, the visual filter is the primary engine to disseminate the meaning, and the importance of sight becomes axiomatic. To this it is added an intelligent interpretation based on the sensorial experience, on memory and on theories and specific methods assimilated from the already existing written sources. All these together enable the formation of a minimum vocabulary necessary for exploring the meaning of landscape. In this respect, Paul Groth stated that, despite the fact that J.B. Jackson's studies found a unanimous acceptance among the United States geographer and not only, no single paradigm took shape to control the cultural landscape studies in the first half of the 20th century[18] (Groth, 1997, p.3).

The cultural landscape opens before us like a book, ready to be read. Reading this book means first of all watching. But, as we have stated before, we must not narrow everything down only to sight as sense. This thing points out precisely the main characteristic of the landscape reading process, which implies filtering the information by means of the specialized filter of the intellect, irrespective of the senses we use[19] (Spirn, 1998, p.36). The experience of landscape reading is a multi-sensorial one, and the individual uses his/ her five senses. As Yi-Fu Tuan states[20] (Tuan, 1977, p.10), there is a close connection between seeing and thinking, which makes these two processes to be practically correlated and essential in the process of interpreting, understanding and reading the landscape. But it is not enough. Reading the landscape is a much more complex process. At the beginning of the 90s of 20th century, other theo-

---

[18] The original quote: "In spite of Jackson's centrality, no single paradigm has controlled cultural landscape studies in its first half-century. Two generations of writers and scholars have added their own questions, types of sources, and traditions into cultural landscape study, and the parameters of the enterprise are still open. The following statements enunciate very basic but widely held tenets of cultural landscape studies in the 1990s".

[19] Anne Whiston Spirn shows us how John Hull, a person with sight disability, perceives the space by means of hearing . "In Touching the rock, John Hull, who is blind, describes relying on his hearing for spatial perception and orientation".

[20] The original quote: "To see and to think are closely related processes. (…) Seeing, it has long been recognized, is not the simple recording of light stimuli; it is a selective and creative process in which environmental stimuli are organized into flowing structures that provide signs meaningful to the purposive organism".

ries came up, and the notion of text was extended to the one of intertextuality. Therefore, in the method of observing and reading landscape as a text, it is accepted the theory of the pre-existence of a written material that is supposed to have influenced what is now deciphered as landscape textuality. As a post-modern response, the focus was directed on the complexity of images represented by landscape and on the "text" enciphered in it. Thus, a landscape can be the subject of several interpretations influenced by the cultural position of the observer or it might be simultaneously interpreted in several ways.

In the professional literature, we often meet the discussion around the issue of a specific language for landscape. We cannot approach this topic without making reference to Anne Whiston Spirn's study[21], which represents a reference work in the field. For the beginning, we choose not to analyze the text itself, but rather the way in which the book is presented. At its beginning, there are three quotations belonging to an architect, an environmental historian and a scholar of literature, quotations meant to be critical interventions on the purpose and approach of the study. The three persons mentioned reason on the title: why using the term of landscape associated to the one of language? The architect criticizes the use of the term landscape, because the way in which the author approaches the topic implies using the terms city or building in various contexts. The scholar comments on the fact that landscape is not language, but merely representation, and the historian wonders why the author refers to landscape, when she could have used terms such as environment or place, as the word landscape is too ambiguous[22] (Spirn, 1998, p.1).

From our point of view, Anne Whiston Spirn gives an answer to these questions in the very first pages of her book. The author justifies her choice of the term landscape, precisely because it has been (and still is) the environment in which humans have existed from the beginning, in nature, next to animals and plants. Therefore, landscape has become the protective environment which offered them everything they needed for survival. The language of landscape has thus become the native language of humans[23] (Spirn, 1998, p. 15). The language of landscape is a language that is no stranger to any individual, but

---

[21] The work is rich in information the author gathered during the 12 years it took her to write it, being perhaps the most extensive one of this kind. Along the years, it has received the unanimous appreciation of specialists, among which Yi-Fu Tuan is certainly worth mentioning, an author who marked the thinking of American geographers in the second part of 20th century.

[22] Following the table of contents of Spirn's book, there are three quotations of anonymous authors. They will be mentioned here, for a better understanding of our discourse : "Why the language of landscape? The language you describe is about so much more than landscape, I includes buildings and cities. [An architect] Why not the language of environment, or place? Landscape is so ambiguous. [An environmental historian] Landscape can't be language! You mean landscape is representation. [A scholar of literature]".

[23] The original quote: "The language of landscape is our native language. Landscape was the original dwelling; humans evolved among plants and animals, under the sky, upon the earth, near water. Everybody carries that legacy in body and mind. Humans touched, saw, heard, smelled, tasted, lived in, and shaped landscapes before the species had words to described what it did. Landscape was the first human text, read before the invention of other signs and symbols".

the levels of understanding and "speaking" are various. Every living being has its own "device" to decode and read the landscape. The same way, each one has the capacity to interact with the landscape, but on different levels. No individual will have a coherent image on the whole which would represent an absolute truth. Each of us will have a different image on the same landscape.

Moreover, there is a close connection of the landscape language to the human language, a fact that was also stated by Christian Norberg-Schulz. He makes a connection between the names of places in human language and their physical existence. The fact that the names of places are nouns, the author states, is the very proof that they are entities that exist[24] (Norberg-Schulz, 1979, p.16). Space, however, as it is a system of relations, is associated to prepositions that determine its topological character[25] (Norberg-Schulz, 1979, p.16). Along the same line, places represent dialects of a language. For Anne Whiston Spirn these dialects are born from the enduring dialogues the individual carries with space[26] (Spirn, 1998, p.181).

It is worth mentioning that improving such language means a great effort that enable the assimilation of some terms and procedures that should become ordinary within a daily used language. It is the same as when one learns any new foreign language. It is a difficult process and the later it is done, the more difficult it is to assimilate. The language must hence become a reflex, so there should be no reserves or babbling, and this can only be obtained by constant exercise.

Thus we have a paradox[27] (Popescu, 2001, p.16.), in the context in which the lack of constructive rigidity of the cultural landscape studies is notorious, as they are flexible in terms of adopting various principles and theories in the evolution process. At the same time, the process of investigating and identifying cultural landscape is sometimes uncertain, ambiguous, subject to some subjective approaches. It may have a speculative character, based on an exacerbation of the landscape symbolism[28] (Lewis, 1979, p. 23), but it is important to

[24] The original quote: "Places are hence designated by nouns. This implies that they are considered real 'things that exist', which is the original meaning of the word 'substantive'".

[25] The original quote: "In our daily life we hardly talk about 'space', but about things that are 'over' or 'under', 'before' or 'behind' each other, or we use prepositions such as 'at', 'in', 'within', 'on', 'upon', 'to', 'from', 'along', 'next'. All these prepositions denote topological relations of the kind mentioned before".

[26] The original quote: "Local landscape dialects emerge out of dialogue with enduring contexts of place; traditional vernacular landscapes are a consequence of collective learning, trial and error, finding what works and repeating it, refining through experience".

[27] The well-known aesthetician and literary critic dr. Titu Popescu stated: "In aesthetics, paradox means the apparent solution of an enigmatic situation (a process which leads to the satisfaction of a relaxation) and the emotional agent is the unexpected, the unpredictable (which generates a disturbance and a new tension). Therefore, paradox is at the same time conclusion and challenge, representing a synchronism of the contrary aspects, lending a true specificity to it".

[28] The original quote: "For example, we can speculate endlessly (and often pointlessly) about the "symbolism" of, say, the American front lawn, made of mown green grass: perhaps it is a status symbol reflecting a borrowing from England, and thus a subliminal reflection of our admiration

know the fact that the approach of cultural landscape is dichotomical with a duplicity character. Thus, the scientific approach must be both phenomenological and semiotic, on two levels, a descriptive one and an interpretative one.

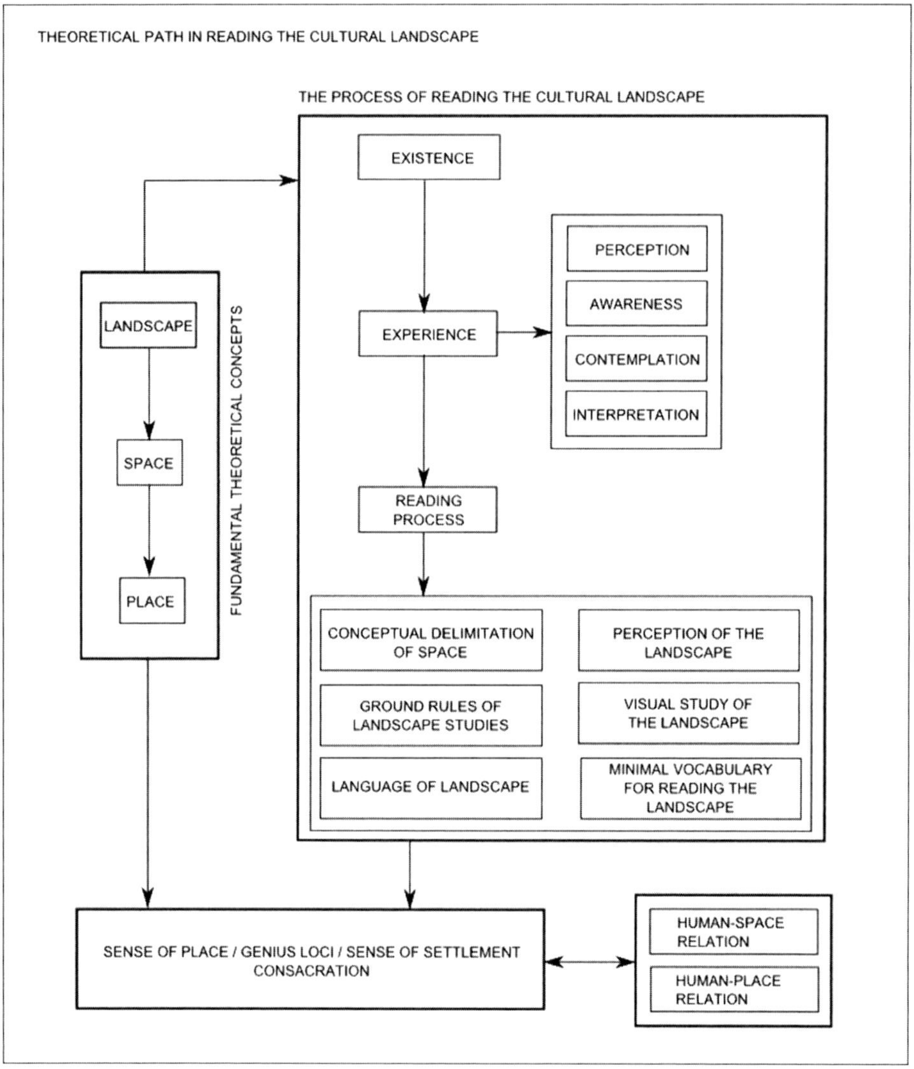

Figure 5. Process of reading the landscape. Copyright: Alexandru Calcatinge.

for things English. But much of that "symbolic speculation" is likely to be hot air unless we really know how a lawn works in a very mechanical way".

7.1.3 Axioms for defining the Landscape

Peirce F. Lewis presented in 1976[29] seven axioms of the process of reading the cultural landscape, each having a series of corollary that the well-known American cultural geographer considered to be important and directly connected to it. Knowing these axioms as well as the study context of cultural landscape is important, all the more as the school of town planning, landscape architecture and territory development ensures the training of those who are space creators and planners par excellence. Lewis's axiomatic principles are based on a simple attribute according to which all landscapes have a cultural meaning[30] (Lewis, 1979, p. 12). And here is how, more than thirty years ago, the American geographer was confronted with a misunderstanding and a disapproval of the common individual towards cultural landscape, and he attributed most of the blame to the academic disciplines[31] (Lewis, 1979, p. 13) at that time, which did not teach their students the importance of knowing and understanding landscape in order to "decipher" as well as possible the things that surround them.

We will thus continue to briefly present the axioms mentions above, trying to accommodate a clear and concise understanding. The first axiom makes reference to landscape as clue to culture (the axiom of landscape as clue to culture) (Lewis, 1979, p. 15) or, as it is briefly presented by the author „the culture of any nation is unintentionally reflected in its ordinary vernacular landscape"[32] (Lewis, 1979, p.15). There are five corollaries associated to this axiom. The first corollary is the one of cultural change (corollary of cultural change) and it makes reference to the tight relation between cultural landscape and national culture[33] (Lewis, 1979, p.15). Thus the author states that any major change of the cultural landscape involving a sustained effort will definitely be connected to a possible change in the national culture. The second corollary (the regional corollary), refers to the way in which two regions of a territory or town can look different on the national territory as a result of the cultural differences of the individuals living in these regions. The third corollary (corollary of convergence), the one of cultural convergence deals with the way in which two regions may have similarities at the level of cultural landscape as a result of the

[29] Lewis's axioms were published for the first time in 1976 in Journal of Architectural Education, but in our work we will refer to the version published in 1979 in The Interpretation of Ordinary Landscapes published by D. W. Meinig.

[30] The original quote: Before presenting the axioms, Peirce F. Lewis stated: "The basic principle is this: that all human landscape has cultural meaning, no matter how ordinary that landscape may be".

[31] Lewis stated: "very few academic disciplines teach their students how to read landscapes, or encourage them to try".

[32] Lewis stated: "In other words, the culture of any nation is unintentionally reflected in its ordinary vernacular landscape".

[33] The original quote: "We must conclude that if there is really major change in the look of the cultural landscape, then there is very likely a major change occurring in our national culture at the same time".

convergence of the cultures of the communities living in them. The forth corollary, the corollary of diffusion refers to imitating the cultural landscape in a region in another region as a consequence of the way in which the material culture was spread by the population's migration flows. The fifth corollary, the corollary of taste, speaks about the tastes the different cultures have with respect to the cultural landscape[34] (Lewis, 1979, p. 15).

The second axiom refers to cultural unity and landscape equality, claiming the fact that each element in the landscape has a cultural meaning. The lesson taught by the author after stating this axiom is that there is no such thing as uninteresting landscape from the cultural point of view, irrespective of how ordinary it may be[35] (Lewis, 1979, p. 19). This axiom has no corollary, but there are three warnings presented which draw our attention on the way in which the unique elements in a site must be looked upon, but also the plain, ordinary ones, which tend to be neglected during the process of identifying and investigating the cultural landscape, precisely because of the simplicity and habitualness it represents.

And with this, the author goes on to the third axiom, the axiom of common things. This one refers merely to the toilsome way in which the study of ordinary landscape is done at academic level, associating it to negligence and snobbism. This axiom also has a corollary, the corollary of nonacademic literature, of literate nonacademic professional literature. By this corollary, the author explains how the necessity of writing about ordinary places and cultural landscapes as well as the adopted literary style is closely connected to the place in which the individual will try to gather evidence and he provided some examples. Among these, he mentions various journals which deal with different topics dedicated to the authentic cultural landscape, written in an essay manner, accessible to anyone, but each one having a clearly defined target audience, and not tabloids. He mentions these not because of the way in which the articles are dealt with, but especially because of their variety, which is different from the emblematic landscapes of the symbol buildings which are always mentioned by the academic publications. The author explains himself and tries to remind the fact that the older, consecrated journals are the ones one must give credit to, presenting, early for that time, a reality nonexistent at us[36] (Lewis, 1979, p. 21).

The fourth axiom is the historic axiom, stating the fact that during the process of understanding the contemporary cultural landscape, history plays a major part. The reality surrounding us now is the one that was built in the past, by

[34] The original quote: "A huge amount of our day-to-day behavior and the landscapes created by that behavior is dictated by the vagaries of "fashion" or "taste" or "fad". And when we speak of "taste", we are talking about culture – not about practicality".

[35] The original quote: "the moral is plain: no matter how ordinary it may seem, there's no such thing as a culturally uninteresting landscape".

[36] The original quote: " Trade journals, especially the old, established ones, can usually be trusted for their cultural judgments. (If their appraisals are wrong too often, they simply do not stay in business)".

individuals who had other training, other needs, other tastes, other types of technologies or ambitions. What we have today is part of the vision of those before us. Our present is what future meant for them, the need to build and construct for future generations. What may be old-fashioned today, or inefficient, in their present it was useful, pleasant and opportune. For this axiom, the author defines two corollaries. The first one is the corollary of a historic lumpiness, supporting the thesis according to which the cultural changes never took place gradually, but they were always influenced by a major historic event[37] (Lewis, 1979, p. 23). The second corollary is the mechanical (or technological) one, referring to the way in which one can perceive the cultural meaning of a landscape by knowing the technology used in creating certain elements under study.

The fifth axiom is the geographical (or ecological) one, according to which the cultural landscape elements lose their meaning if they are studies out of their geographical context[38] (Lewis, 1979, p. 24).

The sixth axiom refers to the control on the environment (axiom of environmental control), mentioning the way in which cultural landscape is strictly connected to the environment, and understating the elements of the former implies the comprehension of the physical environment.

The last axiom, the seventh, refers to landscape obscurity, to the way in which the landscape is the one which sends messages difficult to understand. For the author, everything is about the way in which the questions related to landscape are asked. The capacity to answer these questions and to look for answers in the right places requires a particular intelligence. The author suggests the alternation of thinking about landscape with its contemplation and reading, in repetitive cycles, which can clarify the way in which we perceive it, identify and interpret it.

One can easily observe that these axioms are applicable even thirty two years after their first publication, in a continuously changing world and in the middle of the globalization process. This thing leads to a perpetual attempt to understand what cultural landscape is actually, to decipher the meanings and symbols of an environment always subject to transformations and changes.

However, before analyzing these frequent principles in the cultural landscape study and presenting new ones, it is highly important to know the context in which they appeared. As a consequence, the academic discourse slowly turned into a manifest for knowing and understanding the concept by the general public. Thus one attempted to exemplify the terms used in this respect in a way that would enable the uninitiated individuals to acquire more easily the notions that

[37] The original quote: "Most major cultural change does not occur gradually, but instead in great sudden historic leaps, commonly provoked by such great events as wars, depressions, and major inventions. After these leaps, landscape is likely to look very different than it did before".

[38] The original quote: "This axiom is so obvious that it should not have to be mentioned, except that so many scholars and "practical" people persistently flout it".

would help them understand the cultural landscape[39] (Lewis, 1979, p. 14). We can very clearly see why this reaction of the American geographers towards the individuals' lack of interest for landscape, by perusing the volume edited by Victor Bondi with the title American Decades 1970-1979. He considered, in extremis, that the decade of the 70s did not exist[40] (Bondi, 1995, p. VII), being somewhere in the shade of the previous decade.

Of course, in the support of the argument, one must not forget the fact that we deal with what happened on the United States territory, as it was here where the pulse was felt the most of what the studies of cultural landscape represented in the second half of the 20th century. In Europe, however, we can associate the British geographer's doctrines to the ones in the United States, remaining still in the context of a strong Anglo –Saxon influence on the development of the cultural geography. Nevertheless, the decade of the 70s was full of changes. The United States were in a serious ideological crisis, in which the society's need of change was based on consumerism, economic evolution, fight against communist doctrines and the end of liberalism typical for the preceding decade. It was a period of conspiracies, of the big corruption cases (the well-known Watergate example), of paranoia created by a vulgar drug-addicted world, of the social and energetic crisis[41] (Bondi, 1995, p. 128). The new ideologies thus led to geopolitical changes, and the price for keeping the American life style was difficult to bear by the society[42] (Bondi, 1995, p. XI).

---

[39] Lewis stated the following related to the cultural landscape studies and to the way in which they were perceived by students: "Years ago, when I started teaching about cultural landscapes of the United States, I was puzzled and annoyed that students seemed so obtuse. They seemed blind to all that marvellous material around them, and even worse, some of them seemed insulted when they were told to go outdoors and use their eyes and think about what they saw. Gradually, I realized that the students were not obtuse; I was. The students were simply aping the great majority of their faculty mentors, who by their inattention to ordinary landscape were teaching the students very effectively that landscape didn't matter: that serious students did not deal with trivial questions about ordinary everyday things".

[40] In the introduction of the volume mentioned in the text, Victor Bondi stated: "It is easy to dismiss the 1970s as the decade that never happened. The political and cultural trends of the 1960s continued to dominate life in the United States at least until President Richard Nixon's resignation in 1974; the political and cultural trends of the 1980s were manifested with increasing visibility for the remainder of the decade. (…) The 1970s, it seems, have little to define them except, perhaps, their nothingness".

[41] The original quote: "A major source of the instability and change of the economy during the 1970s was energy. In early 1973 the United States faced shortages of electricity, gasoline, and heating oil, leading to the shutdown of factories and schools, the cancellation of some commercial airline flights, electrical brownouts, and lines at gasoline service stations. Blackouts plagued cities and industries, most spectacularly in New York City on 13- 14 July 1977. High fuel prices reduced the productivity of American industry. Heavy imports of fuel harmed the U.S. balance of payments and destabilized the international monetary system".

[42] The original quote: "Because the 1970s were the crisis decade for established American political, economic, and social assumptions, it was also a decade wherein Americans attempted to construct new, more realistic, ideologies. The exposure of government misdeeds exemplified the willingness of some political figures to resist the more brutal effects of governmental power and demonstrated their confidence that U.S. institutions were capable of honesty and reform. (…)

The Europeans had the situationsts (Situationist International), situations adepts and creators, willing to discover and experiment the urban environment, who had their roots in the surrealistic doctrines of the 20s. As for the Americans, the evolution of the concept of cultural landscape from the definition given by Carl Ortwin Sauer showed the public's lack of perception in terms of the way in which cultural landscape can be deciphered and observed without the necessity of a specialized training. All you needed was an increased attention on what surrounded you. This type of landscape interpretation was no news for the geographers such as Peirce Lewis or D.W. Meinig. Concurrently, John Brinckerhoff Jackson also delivered a series of lectures and published several books[43] and articles related to the interpretation of landscape, its discovery and understanding, being a promoter of the methods suggested by theoreticians in the second half of the last century.

7.1.4 Donald W. Meinig's basic principles in defining Landscape

In order to support Pierce Lewis's presentation of axioms exposed above, we must mention D.W. Meinig's description of the ten ways in which landscape can be perceived. With an apparent simplicity, as Lewis's axioms, D.W. Meinig presents ten possible versions of landscape, as they can be perceived by the individual. The similitude between the two approaches manages to point out the evolution of the concept of cultural landscape. With the time, it has proved to be solid and able to assimilate new components, which would contribute both to an increased understanding and to an inevitable usage of the operative concept in various research fields. At the same time, this fact has led to a poor approach in some branches related to them.

Geographer D.W. Meinig promotes the thesis according to which, in landscape interpretation, individuals can perceive at least ten hypostases of the same landscape. Here is their brief presentation. First of all, there will be individuals who will perceive the landscape as nature. They are the ones who, par excellence,

---

These experiments make the 1970s a rewarding period for historical study, but they also dramatically reveal the limits placed by the public on institutional and ideological innovation, because all of these innovations failed. They failed primarily because the American public did not embrace them, preferring to support established institutions".

[43] J. B. Jackson is the author of several reference books in the studies of cultural landscape, both worldwide and especially in the United States. During his training, he was influenced by the works of Lewis Mumford and Oswald Spengler, which helped him in his attempt to "derusticate" the concept of cultural landscape introduced Carl Ortwin Sauer. After his death, in 1996, the American Geographers Association proposed the Jackson prize, taking the name of the prominent geographer, in order to award the American geographers. Among his books, the most popular ones are Discovering the Vernacular Landscape, A sense of Place, a sense of Time, Landscape in Sight and Looking at America.

will mentally annul man's intervention from the landscape. There are nostalgic persons who will bring their landscape interpretation to its initial unaltered condition[44] (Meinig, 1979, p. 34). It is a method to revive romance, nostalgia and seduction in landscape, a method known two centuries ago as the only contemplation method. It is a necessity, and the annulment of any intervention may bring forth for the one who watches it a side of the landscape which otherwise would be difficult to observe. For others, landscape can be seen as habitat as well. For those in this category, landscape can also be interpreted as the individual's shelter, a space in which he happily connects to nature. It's a space of the individual's day-to-day life, and daily work[45] (Meinig, 1979, p. 36). However, there are persons who think of landscape as an artifact[46] (Meinig, 1979, p. 36). These are the ones who, before observing nature, notice man's intervention on it. Nature is merely a scene in which man shows off his craftsmanship in the building domain. Ideologically, Meinig associates this thing to man's condition of creator on nature.

The scientific motivation is the one which determines the desire of acquiring knowledge, in the author's vision. Thus, landscape is presented to us as a system. The one who associates landscape to a set of systems is the one who sees beyond the mere presence of an element in the landscape. These are especially the specialists, for whom, depending on their specialization, nature appears by means of their knowledge. Nature is a study laboratory, a place where scientific experiences occurnad which involves a cultural characteristic[47] (Meinig, 1979, p. 38). Then, there are the ones who see landscape as a problem. However, D.W. Meinig makes a clarification here, explaining that he does not refer to a scientific problem, but rather to something that needs correcting. In the author's opinion, the individuals in this category focus on the environment at they should have a little of the other categories mentioned above[48] (Meinig, 1979, p.

---

[44] The original quote: "Such a viewer is ever tempted in his mind's eye to remove man from the scene, to restore nature to her pristine condition, to reclothe the hills with the primeval forest, clear off the settlements, heal the wounds and mend the natural fabric – to imagine what the area is really like".

[45] The original quote : "Every landscape is therefore basically a blend of man and nature. Man may make mistakes, damage nature and thereby himself, but in the long run man learns and nature heals".

[46] The original quote: "Nature is fundamental only in a simple literal sense: nature provides a stage. The earth is a platform, but all thereon is furnished with man's effects so extensively that you cannot find a scrap of pristine nature".

[47] The original quote: "Such a view is wholly the product of science, a means of looking inside matter to understand things not apparent to the naked untrained eye. (…) For such persons the landscape that others may see is only a facade which their vision penetrates to reveal a transect of intricate pulsating networks, flows, interactions, an immense input-output matrix".

[48] The original quote: "Nevertheless, this view of landscape through the eyes of the social actionist may incorporate something from all these other views: it evokes a reverence for nature, a deeply felt concern for the health as habitat, and a conviction that we have the scientific ability to right these wrongs. What is needed is a far greater awareness of what is happening and why. It is thus a view which tends toward a humanism harnessed to politics in the hope of generating a

40). Afterwards he presents the category of those who consider landscape as wealth. These are the persons who view everything through the perspective of money and market economy, and thus each element in the landscape is given an economical value[49] (Meinig, 1979, p. 41). Landscape as ideology is perceived by those who see beyond its façade, the ones who associate symbols and hidden values to a philosophy or to a culture. We will present the author's entire description in the footnote, for its clarity and in order to enable its understanding[50] (Meinig, 1979, p. 42). Anne Whiston Spirn also states that the landscape we observe is an ideological product which gives birth to a series of essential questions regarding the way in which we perceive it, we understand it and we model it[51] (Spirn, 1998, p.244). As the way to perceive the cultural landscape in terms of ideology is the very way to think of how it was created, Meinig goes on and describes the way in which landscape is seen as history. It is the witness of the changes occurred along the years, being the very scene of these changes[52] (Meinig, 1979, p.43). Everything that man created was left as a proof inscribed in the landscape. This is like an immense data base which records in-formation on each and every element. Thus we get to the landscape as a place, a well-defined space with special features, which includes a little of each category mentioned so far. It is a landscape which cannot be generalized as usual and can be a substitute of a specific, clear interpretation to the point. This is the way to perceive landscape which is most specific for geographers, for which the study of the place and of the characteristics of a space were the cen-

genuine populist movement against what is regarded as a callous, selfish, or simply inert establishment".

[49] The original quote: "Public properties – schools, libraries, streets, parks, reservoirs, garage dumps – are fitted into the system, for each affects the value of its surroundings, as do other site qualities – trees, hills, valleys, and especially views from residential property. Furthermore, people have a place in such an appraisal, for where the rich and where the poor live, work, shop, play, and go to school greatly affect property value".

[50] The original quote: "Just as the scientist looks through the facade of obvious elements and sees processes in operation, so others may see those same elements as clues and the hole scene as a symbol of the values, the governing ideas, the underlaying philosophies of a culture. Where those who see landscape as problem see disorder, clutter, incongruity, congestion, pollution, sprawl, and dereliction amid the glitter, those who see it as ideology may see distinct manifestations of American interpretations of freedom, individualism, competition, utility, power, modernity, expansion, progress. That does not mean they cannot see the problems, but that they are more concerned to look more deeply to see how the landscape represents a translation of philosophy into tangible features".

[51] The original quote: "Competing ideologies govern the ways landscapes are shaped. Ideologies based on loyalty to a primary authority set the terms of dialogues: the authority of nature, an appeal to the natural; the authority of the past, an appeal to tradition; the authority of art, an appeal to invention; the authority of expedience and efficient function, an appeal to economics and the practical, the authority of power, an assertion of strength, force or wealth".

[52] The original quote: "The visible landscape is not a full record of history, but it will yield to diligence and inference a great deal more than meats the casual eye. The historian becomes a skilled detective reconstructing from all sorts of bits and pieces the patterns of the past".

tral themes along several decades[53] (Meinig, 1979, p. 45). They have a constant importance in a century when values and ideologies are shaken by social, economic, global crises and new national doctrines which lead to changes in mentality and perception on the way in which we relate to the landscape and the environment. The last way to perceive landscape described by the author is the aesthetic one. This way to look at landscape is perhaps the most debated one, the most subjective, speculative, sometimes even touching the derisory. It is the mode that triggers the most divergent reactions from specialists[54] (Meinig, 1979, p. 46).

7.2 Understanding the concept of Culture through consciousness, specificity and local identity

In completing the concept of landscape, we must also establish the way in which the term of culture is understood within the phrase of cultural landscape. This will enable us to outline a definition with which we will operate throughout this work, or even more, to make out why culture plays a crucial role in understanding the concept of cultural landscape.

According to the Explanatory Dictionary of Romanian Language "cultura" (culture) means "the totality of material and spiritual values created by mankind"[55]. For our study, however, we will appeal to a broader description given by Tudor Vianu (1982) to the term of culture. Thus, the renowned Romanian philosopher stated: "the cultural will under the double aspect of the purely volitive element which represents it and which consists of energy and inner productivity; the optimistic faith in the themes that continuously challenge mankind ; the value to guide this energy; the cultural asset which embodies this value; the subjective cultural act which cherishes the cultural assets; the objective cultural act which embodies values in a material: here are the elements that build up together the content of the notion of culture" (Vianu, 1982, pp. 158 – 159). Acording to the same author, there are two conditions of culture (Vianu, 1982,

---

53 The original quote: "In this view every landscape is a locality, an individual piece in the infinitely varied mosaic of the earth. (…) It is landscape as environment, embracing all that we live amidst, and thus it cultivates a sensitivity to detail, to texture, color, all the nuances of visual relationships, and more, for environment engages all of our senses, the sounds and smells and ineffable feel of a place as well".

54 The original quote: "Artistic quality is of course a matter of endless controversy. (…) The very idea of landscape as scenery is a surprisingly late development in Western culture, requiring as it does a special conscious detachment by the observer. (…) This, too, is a penetrating view. It seeks a meaning which is not explicit in the ordinary forms. It rests upon the belief that there is something close to the essence, to beauty and truth, in the landscape".

55 According to DEX online. : http://www.dex.ro/cultur%C4%83. accessed on 20th August 2010 at 16:05.

pp. 224 – 259). Presenting them briefly, one is the material condition and the second is the spiritual condition. These two conditions belong to the material theory of culture mentioned by Tudor Vianu and they are derived from the theory of the environment, the economic conditioning and the theory of races as elements of the material condition and from tradition and quantity of the social group as elements of the spiritual condition.

In the same line of thoughts, Alexandru Tănase (1968) in his cultural studies determined that it is important to make a distinction between the concepts of material culture and spiritual culture. This because there exists a contrast between the human and the material aspects of culture. Thus, he determines some constitutive moments of culture which are knowledge (the gnoseological moment of culture), value (the axiological moment of culture), creation and social generalization and critical assimilation of culture (Tănase, 1968, p.89).

As values are objects of consciousness (Vianu, 1982, p.72) they belong to a certain structure. Thus, values are involved in almost every act of knowledge. This act, in its turn, is based on an existing level in the development of the society. In this way, every object from our world becomes full of meaning and a mediator of relations, and the value becomes "that dynamic virtue by which possibility is updated and experience is achieved" (Tănase, 1968, p. 106). Therefore, "Value does not belong to natural things, irrespective of human activity and thinking. (…) In the modern era, the primacy of human activities over nature, the evolution of science, the successes of the action to overpower nature have radically changed the perspective of the human relations to the totality of the real; (…) It is not the being, but knowledge and value that are in the center of discussion, as the subject always tends to impose its seal on existence. But becoming the masters of nature is a moment which dialectically denies, by overcoming it, the moment of the pure and simple agreement with nature" (Tănase, 1968, p. 107).

7.2.1 Understanding Culture through human consciousness and values

Human's account with nature, through consciousness and values, determines oneself to be the master of the latter through the act of culture, which is, as we will see, determined by consciousness. When saying the word consciousness one surely thinks about mind and brain, about human beings and their experiences and their reactions. Therefore, consciousness, apart from its psychological or medical (neurological) definitions and importance, could be defined by terms like awareness or perception, and it is applicable only to biological organisms. A brick does not have a consciousness, but it exists in our consciousness as human beings. The same is for any construction or manmade material or natural element like trees or plants, even though the evolution of science and computers in the 21st century could end to a virtual machine consciousness.

As William Seager stated (1999) the informational value of a neural system of a brain is determined by four sources. Thus, "the first is the link to motor pathways which facilitate the appropriate response to the object which they represent. (...) The second informational role also involves memory, but more broadly conceived. The mind can reactivate these representations to retrieve sensory information. (...) These brain representations also serve, third, as the source of imagination, which according to Descartes requires a 'corporeal figure' for the mind to contemplate. (...) A fourth information function is the production of conscious experience itself "(Seager, 1999, pp.8-9). Thus the four ways to produce informational value as described earlier could be associated with the geographical region and thus the consciousness could be achieved, as William Seager argued, on a two layered way of representation "the bottom layer being representations of combinations of sensory qualities, the higher layer being the representation of cognitively rich content, the prime examples of which are simply the ideas constituting the states of consciousness involved in our normal intercourse with the world" (Seager, 1999, p.9).

Saying this, it is inevitable for us not to remember Yi-Fu Tuan and its arguments on experience as the very engine of life, action and thought. This matter had been studied by us in another paper on the process of reading the cultural landscape. Furthermore, the experience is closely related to perception and thus consciousness, as argued by William Seager, presupposes a sensory perception and a cognitive perception of the world. On a shorter statement it is about feeling and thought.

Consciousness is, in a simple manner, besides being of different kinds, a dualism of subjective and objective which in fact could be substituted by the term awareness. We will further demonstrate this, as the primary focus "on what has become perhaps the central issue in consciousness studies, which is the problem of integrating subjectivity into the scientific view of the world" (Seager, 1999, p.10) that is the objective. If consciousness is awareness as we stated before, then it presupposes certain knowledge based on the constructed reality around us. This knowledge is based on an experience that itself is based on feelings and thoughts (perception), thus on existence. The result is that knowledge is actually existence, and existence is based on a duality. There is a material and a spiritual existence which in a basic term are subjective, but in a world of objective realism. To continue, it would mean that awareness is existence and that it is a dualism itself.

Furthermore, an interesting view for our demonstration is that "all we could possibly have knowledge of are constructions out of subjective experience and that any distinction between reality and appearance within the realm of perception and scientific investigation would have to be based upon some set of relations holding amongst our experiences. He [Kant] added the remarkable idea that these relations were a reflection of the structure of the mind itself – concepts of space, time, and causation are necessary conditions for the existence of experience of an 'external world' and are 'discovered' in that world because they pre-exist in the mind. (...) Kant called this style of reasoning transcendental argumentation" (Seager, 1999, p.23). Thus, "paradigmatic examples of consciousness are the perceptual states of seeing and hearing, but the nature of the consciousness involved is actually complex and far from clear" (Seager, 1999, pp.9-10). Furthermore, the same author William Seager states that "philosophers argue about whether all thinking is accompanied by or perhaps even constituted out of sensory materials (images have been the traditional favorite candidate material), and some champion the idea of a pure thought-consciousness independent of sensory components. In any event, there is no doubt that thought is something that often happens consciously and is in some way different from perception, sensation, or other forms of consciousness" (Seager, 1999, p.10).

Dualism in consciousness studies has been also a key aspect since Descartes. For him there was a substance dualism, "(where a substance is something that can in principle exist all by itself). One is the extended substance, or matter; the other is the thinking substance, or mind" (Kriegel, 2007, p.39). There is also a property dualism in consciousness studies, a modern dualism where "the idea is that even though there is only one kind of stuff or substance, there are two kinds of properties, mental and physical, and neither can be reduced to the other" (Kriegel, 2007, p.39).

This dualism is also underlined by Alexandru Dragomir, who stated that our existence (which itself has a duality) takes place in a dualistic world. Thus, "we live in a double world: a world of nature and of techniques, of lie and reality, of <<democracy>> and political undergrounds, of <<theory>> and of interests, of appearances and of espionage. Everywhere there is a need to <<double>> the world. And science <<doubles>> nature in a way, by technique. (...) Doubling is done every time at the expense of what it is doubled, but, unlike Manichaeism, the double has the meaning of <<the true>> reality, or it becomes something that we don't yet know, and this is: a second reality" (Dragomir, 2008, p.17).

Alexandru Tănase debated the issue a dualism of values and stated that, as "the intellect gives us the representation of the real, an adequate knowledge of it, the will justifies it and modifies it. The two functions of consciousness should act solidarily, but there often appear contradictions between them. For, as the intellect adjusts its operations depending on the properties of the reality, the will models the reality according to its own projects" (Tănase, 1968, p.107 – 108). Therefore we can agree, as Alexandru Tănase does, that in the center of all this values, "the subject – the human being, community and not just will – must take action as an active factor with regard to the natural or human, in order for the jump from nature to culture, from material quality to value, to occur" (Tănase, 1968, p.108). And intellect and will are unitary functions of consciousness and those are the tools that provide us knowledge of the world and the means for its transformation (Tănase, 1968, p.108). Thus, in the context of the theory of values, architecture/ urbanism is "a discipline that must have its own ways to nurture and to express objective and subjective ideas, a discipline that creates permanent values" (Doicescu, 1983, p.40). It is the most tangible and mature form of consciousness. It is the oldest way of existence of a creative act. It [architecture] is the protective shell of our private and [urbanism] of our social life. It gives us the field of expressing all other creative, economical, political, religious or moral acts. It is mean and purpose at the same time. Despite the overall evolution of our times, architecture, at least at a moral and spiritual level, has a specific recurrence. It is the mirror of present and past times and has the power to overcome major shifts in history. Octav Doicescu (1983) stated that "what is left behind by a creator, or even by an ingenious inventor, are the works carrying with them the thinking of his time, the scientific obsessions or hopes of collective or individual outgrowing, everything that preoccupied and preoccupies – was that one, the one he experienced. At most, he has the power to catalyze, in a confession of his spirit, more important thoughts, to emblaze some deeds that pass on without significance in life, to mark out the unity of common confessions, what represents for people their necessities and dignities; what might transform the aesthetic into ethic value" (Doicescu, 1983, p. 102). Thus, architecture has a permanent role in both material and spiritual culture and it is a "creating art, like a spiritual <<alter ego>> of the human being" (Doicescu, 1983, p.25). Therewith, along with the permanent role, there is always a particular one that is given by the creator, the architect, but this means "he must hold a position, think and imagine architecture for the community and build it" (Doicescu, 1983, p.25). In this respect, the same author states that "architecture has remained so far a flower of the spirit, born as specificity of a given human environment and of a given geographical area. With a sharp perception, with all the similar techniques used, it must keep its qualities, which determined it as people's spiritual necessity, as a language typical for the respective culture" (Doicescu, 1983, p. 125).

This particular role that we mentioned earlier determines the specificity and the identity of places and people, as a spiritual and material foundation of life and as a result of cultural processes.

7.2.2 Understanding Culture through specificity and identity

What is the difference between identity and specificity? The answer is the key to determine a certain knowledge of culture in the way we see it in the words cultural landscape. The relationship unity – diversity is contradictory and the very base in the study of identity and specificity, at least in terms of a philosophy of culture. Thus the bipolarity unity – diversity is to be studied with regard to the binomial universal – specific. But what is the difference between identity and specificity? This is the key to determine a certain consciousness of a geographical area. In order to maintain a "local" specificity of our study we will appeal to the writings of Romanian intellectuals on this important subject.

In this respect, in his quest to determine an objective thinking of a human being, the philosopher Ştefan Afloroaei, determined that human thinking is apparently objective, but in fact, as he further demonstrates, is as abstract as it can get. Thus, "it is precisely in his first and elementary thinking that man thinks abstractly, unimaginably abstractly. (…) We think in abstract terms every time when thought retains, with regard to a certain state, only one singular aspect, only one predicate of a subject, only one description of it. Just that, in its rush on the stage of his world, the human being end up seeing only what is visible in the first place and fractionally – nothing else more" (Afloroaei, 2008, pp.131-132). The modern individual is thus a prisoner of its own life, trapped in a high speed chase where space and time are no longer the same as they used to. Everything happens in such a hurry that "human mind can no longer bring together the disparate parts, in order to see how they stay together and grow together, as something real and alive builds up" (Afloroaei, 2008, p. 132).

Figure 6. Mountain villages in Gastein Valley, the Austrian Alps, at the border between Salzburg and Carinthia. Copyright: Alexandru Calcatinge.

Therefore, the study of specificity and identity is not just the perception of the obvious or the concrete and objective. As we live in a discontinued world where cultures are facing the paradigms of globalization, the need to redefine identities and specificity is taking (more and more) ground. Thus the postmodern paradigm determines us to be, as Grigore Georgiu stated, "the witnesses of an almost indistinguishable transition, from a disjunctive paradigm of this rapport – even if it stipulated necessary associations between its terms – to a conjunctive paradigm, for which cultural diversity is organized in a polycentric and horizontal way, not monocentric and pyramidal way" (Georgiu, 1997, p.85).

In this respect, the same author considers that "specificity would be an variable content that fills up some universal alveoli and constants of the human spirit" (Georgiu, 1997, p.83). Identity, however, is seen as a result of cooperation, communication and interplay relations of a culture, the exact opposite of its isolation (Georgiu, 1997, p.87). Therefore, cultural identity is "an existential equipment, technical and symbolical, through which a human community codifies, in a expressive way, its history, existence, reality, life with all its elements. At present, cultural identity is a synthetic marker for the creative acts of a social and ethnic community in the sphere of values, regardless dimensions and historical performances of that community" (Georgiu, 1997, p.97).

An important study on local cultural identity was conducted by the architect Adriana Matei, professor of architectural theory at the Technical University of Cluj-Napoca. In this study, beside the analysis of local cultural identity and local specificity, her interest was also on language, communication, site, place, rural and urban morphology, cultural morphology, local traditions and some other fields related to the urban and architectural phenomena. Thus, from the very beginning she gives a conclusive definition of cultural identity, as it implies a consciousness affiliation to a certain culture (Matei, 2004, p.1) and that it can only refer to one community (Matei, 2004, p.2).

Figure 7. Villages in the Norh-Eastern side of Austria, near Czech Republic. Copyright: Alexandru Calcatinge.

Figure 8. Şura Mare village, in Southern Transylvania. Copyright: Alexandru Calcatinge.

Right after that, she delivers the difference between local identity and local specificity. Therefore, "local identity implies a subjective, consciously assumed pluralism of a community, while local specificity can be provided only by the sum of entities and particular forms that are kindred or identical, thus objective" (Matei, 2004, p.4). Local specificity is represented by powerful characters of a certain area or culture, mostly material, but which are not necessarily sources of creativity (Matei, 2004, p.81). Thus its characteristics are passive, while the ones of the local identity are active. Local specificity is objective, while local identity is subjective and "it implies an acknowledged pluralism" (Matei, 2004, p.81). In her study on local specificity, the author takes a theoretical approach on architectural and urban space, on site and on place and on urban and rural composition.

Lucian Blaga in his well known Trilogy of Culture takes a determinant approach when discussing on the matter of landscape, local consciousness and specificity. Thus, in the study of landscape and unconscious horizon, the Romanian philosopher explains how in the same landscape coexist cultures with different spatial feelings and how in different landscapes coexist cultures with a unique spatial feeling (Blaga, 2011, pp.61-71). In this respect, it is important to note that regarding his theories, the spatial feeling is the kernel of culture, at least as viewed by morphology (Blaga, 2011, p.61). Lucian Blaga states that "the relation between landscape and spatial vision is not so simple and unequivocal as the morphologists imagined it to be. The respective relation involves all sorts of variants, whose significance overruns the one of equivalence every time. Landscape could, nevertheless, integrate by itself, with no frictions, smooth, into a spiritual gear, but at the same time the landscape could not integrate in such a gear, remaining a foreign object, unassimilated, an exterior object of the consciousness" (Blaga, 2011, p.65).

Figure 9. View over the town of Mediaş. Copyright: Alexandru Calcatinge.

Further on, in his attempt to explain the existence of different cultures in the same landscape, he brings into discussion the case of Transylvania (or Ardeal as is also known in Romanian culture) where Saxons and Romanians lived together for hundreds of years in the same geographical area. This way their own spatial visions have managed to coexist and thus give birth to a unique local identity and specificity. Thus, as part of their spatial vision of the Saxons, Lucian Blaga determined that the Gothic spirit, their mystic freedom and a hard-bitten spiritual nature are key components (Blaga, 2011, p.66). When describing the Romanian spatial vision, the author is setting his arguments on the architectural aspects, even though he confesses that many other aspects are equally important and defining, as dancing, traditional writing and poetry and many other popular customs.

Figure 10. A traditional Saxon village in Transylvania. Copyright: Alexandru Calcatinge.

Thus, in comparison with the Saxons houses, which are high and denote the spiritual lift of the people, and with the Russian houses that are developed more on horizontal, the Romanian house is not exclusively high or low, it incorporates a bit of both, in harmony, as the spatial frame of geographical hills and hollows "does not admit the hypertrophy of one tendency and the disadvantage of the other" (Blaga, 2011, p.68). Furthermore, Lucian Blaga considers that the local Romanian spatial vision could be related and akin to the general one in the Balkans, as part of a specificity that has the "determined form of a curled infinity" (Blaga, 2011, p.68). But when he writes about Transylvania, he considers that "the three-dimensional infinity, and the eastern and more vegetative vision of the mioritic space" (Blaga, 2011, p.68) where there live and share the same landscape people that re so different, permit the coexistence of the Saxon and the Romanian in "the atmosphere of eternal ups and downs of its mioritic space" (Blaga, 2011, p.68). Here is how, in the description mentioned above, Lucian Blaga calls the subconscious frame of the Romanian popular culture and life as "spaţiul mioritic" (Blaga, 2011, p.68).

Figure 11. View over the Târnave Region in Transylvania. Copyright: Alexandru Calcatinge.

As the author explained the existence of various cultures within the landscape, he goes on by distinguishing between what he considers to be the minor culture and the major culture. Thus, beside the differentiation on dimensional criterion between the two, Lucian Blaga determines also a differentiation based on a "qualitative –structural" criterion" (Blaga, 2011, p. 332). In the process of determining such a criterion, Blaga mentions the theory of the ages of cultures, tributary to the morphological – biological theory of culture (Blaga, 2011, pp. 332 – 333). It is an extremely interesting and necessary approach for building up a differentiation between the minor culture and the major one. Even if, as the author states, "we assumed the task to issue the death certificate" (Blaga, 2011, p. 333) o the theory mentioned above, the concepts of "childhood" and "maturity" are kept in the construction of the argumentation. Consequently, "the minor culture is not the first age of a culture understood as a self-contained subject, meant to grow up step by step, as it happens with every independent organism. (…) Thus the minor culture is a culture created by the perspective of the man's childish structures and therefore it can last and perpetuate indefinitely" (Blaga, 2011, p. 334). Thereby, "the appearance of a major or minor culture is not a matter of actual age of the creating people. (…) It is said that culture creators have two ages: one is the real age, the biological, individual one – the other is an adoptive age, under the auspices of which they create (…). Hence, the minor culture is a culture created under the domination and constraint of structures that are typical for childhood, in the meaning of << adoptive age>> of creators" (Blaga, 2011, p. 335). We discuss more about the minor culture, for, as we will further observe, Constantin Noica also speaks about us, the Romanians, as being part of a minor culture, which does not mean a qualitative inferiority. Therefore, the meaning of this concept, as it appears in Lucian Blaga's works as well, must be explained more thoroughly. In order to do that, Blaga goes on to build up his argumentation based on the assimilation of the Romanian village t a minor culture. This enables us to have a better understanding of the situation in which our country has been for several decades, as well as of the way in which, in the cultural landscape studies, we would be able to relate to that adoptive age mentioned above or to the fact that in (Blaga's) contemporary philosophy "minor cultures also bear the name of << ethnographic>> cultures, and major cultures are also called << monumental>> cultures" (Blaga, 2011, p. 331). Thus in Blaga's works there appears the issue of differentiating between the city and the village, which represents an increasing preoccupation for our studies on cultural landscape and for the way in which he understand culture in this entire process of analysis and study.

Figure 12. Sibiu (Hermannstadt) at Sunset. Copyright: Alexandru Calcatinge.

In this respect, Blaga states the fact that "the village is not situated in a purely material geography and in the network of space's mechanic features, like the city; for its own consciousness, the village is situated in the center of the world and it is extended in myth" (Blaga, 2011, p. 337). Thus we have already gathered a series of data that are absolutely necessary in order to help building a clear definition of what cultural landscape means, of is theory and of the way in which we can relate to it at rural and urban level. But before we continue our discussion on specificity and consciousness at urban level, we will mention the fact that "lining in the city means living in a fragmentary environment and within the limits set by civilization's rules at every step. Living in the village, however, means living in cosmic sky line and in the consciousness of a destiny shed by eternity" (Blaga, 2011, p.337). In close connection to the two types of living, there are the visions on space and time, which, as we will further see in our work, are considered to be crucial in structuring the concept of cultural landscape. Thus, it is still Lucian Blaga who points out that " the vision on space, in the minor culture, hardly overruns in size the visible horizon; the village, the settlement, the hamlet, is the center of the world, and the world is almost entirely visible from one spot; the horizon that can be seen is extended straight in mythology. In a minor culture, the vision on time hardly overpasses the organic duration of an individual life; beyond this horizon, time is rather something suspended or shapeless. In a major culture, the structures of maturity become sovereign. (...) There starts the guided creation. The plans are no longer available for the individual, but the individuals are subjects to a plan. The person – creator is situated in spatial and temporal horizons which physically overflows the visible enormously. << History>>, as articulated and humped time, is the dimension where a major culture develops and grows" (Blaga, 2011, pp. 345 - 346).

Figure 13. The Romanian village, a childhood's village. Copyright: Alexandru Calcatinge.

The writings of Professor Alexandru Sandu were left to the very end of this chapter for a good reason. They incorporate great theoretical approaches in the field of urban studies in Romania. Moreover, the discussions on identity and local specificity are ones of the most conclusive and complete in the framework of urban studies in our country. As Liviu Ianăşi argued in the foreword of professor's Sandu book, "notions as particularity, character, specific are part of the public discourse on city; they are also exposed – the same way as sustainability or community identity – to over-use or tiredness generated by improper use or manipulative one" (Sandu, 2007a, p.12). What is most important to mention is that the texts of professor Sandu, which consists of five articles, are written in a consecutive manner, from the 70s until the 90s of the 20th century. Regarding this, they may appear as outdated for the post-modern urban paradigms. But, as odd as it may appear, they are actually very actual and fit perfectly in the contemporary urban theories of our country.

A bitter truth about local architectural and urban specificity is revealed by the professor when he states that the relationship between the building and the environment is precarious revealed in a urban space that has resulted from a long spiritual and material existence where different cultures have met (Sandu, 2007a, p.136). The common mistake to reduce specificity to a simple facade or decoration is highlighted as being a wrong approach. Therefore a conclusion would be that "specificity is not a thing in itself and one you try to achieve at any cost; it is the result of a synthesis in time and space, made up through an accumulation of certain traits at the level of the respective structure derived from both spaces' functionality and their ambient value" (Sandu, 2007a, p.141). Furthermore, professor Alexandru Sandu determined a framework for appreciating the urban space quality (compared to the process of appreciating a work of art) and enumerated three key components, as "in assessing the quality of the urban space, beside the effect, a particular concern is also represented by : (1) cause and realization means, upon which – and let us not forget – a direct, objective determination acts through functionality and the possibilities of realization, (2) the adaptation to an existing material that brings out both inventiveness and the assimilation capacity of previous capacity and (3) time as a container and space as a place of topical selection that tends towards an increasingly complex formation of the urban space" (Sandu, 2007a, p.142). As a conclusion, "urban space specificity cannot be a product of details, but a product of the ensemble, the detail helping us just to retain the specificity or it becomes an element of specificity only when, through multiplication, it changes into an ensemble" (Sandu, 2007a, p.149).

Figure 14. Mountain village in România. Copyright: Alexandru Calcatinge.

Writing about cultural specificity in architecture and urbanism, professor Alexandru Sandu states that "national specificity is revealed by its own nature first of all, by something particular to the human community making the nation and its response to the environment in a given situation" (Sandu, 2007a, p.115). Nevertheless the author considers that it [national specificity] is "an outcome of social and historic conditions" (Sandu, 2007a, p.116) and is an intellectually and spiritually compiled and altered architectural forms with the influence of the cultural and local ethnic elements. Thus "the Vienna-based baroque, the German-based Jugendstil or Russian-based classicism are expressed in Banat, Transylvania, Wallachia or Moldova in a autochthony architectural form. This is why we could speak about a characteristic image of the urban space – especially at street level – in many Romanian towns. Although it benefits from a large array of styles, this image neither reproduces nor transposes them but interprets them in a folk or cult manner reflecting some spiritual potential, material needs or capacity of building" (Sandu, 2007a, p.118). In this respect we will appeal to the writings of Alexandru Dragomir who stated that "rural culture (so called) folklore is ageless, the urban culture is evolutionary" (Dragomir, 2008, p.100).

Figure 15. The small town of Hallstatt in Austria is on UNESCO's cultural patrimony list. Copyright: Alexandru Calcatinge.

To complete what professor Sandu argued, we return once more to the writings of Grigore Georgiu who stated that the national cultures are stages of internal contradictions and polarities but it does not mean that the general identity of a national culture is lost. Every culture has a capacity to self adjust and this is bound with the diversity and heterogeneous codes inside it (Georgiu, 1997, p.100). Talking about codes of a certain culture, Grigore Georgiu states that "through their symbolic function, of expressions of a way of life, of na existential style, cultural forms permanently interpret and codify national life in its totality" (Georgiu, 1997, p.101). This is a way to cheat time by transforming the creative force and spiritual values into durable and symbolic values. "Those are the most transparent forms of national specificity" (Georgiu, 1997, p.101). Grigore Georgiu states, and the most powerful codes of culture and national specificity were determined by authors that always expressed the primordial and vital specificity like Eminescu, Creangă, Rebreanu, Blaga and others (Georgiu, 1997, p.101).

As an ending conclusion on this matter of consciousness and local specificity and identity we will appeal to what Constantin Noica once said: "we know that we are what is called <<a minor culture>>. We also know that this does not necessarily mean a qualitative inferiority. Our popular culture, though minor, has track records qualitatively comparable with those of the great cultures. And we know that we have in our popular culture a continuity that the greater ones do not have" (Noica, 2010, pp.7-8). In addition, we can add what professor Alexandru Sandu said with regard to specificity and continuity: "this is why the idea of space and time continuity is so dear to me; that is the awareness that a city cannot live only on its history (the renovation of the old) – something new has to be built. (...) In fact, the difficult thing in our profession resides in our correct understanding of the amount of freedom the place would allow and use it to the benefit of the place, not against it; by the modesty of reproducing a basic addition in stylistic unit or by the refinement of a convenience in which contrast can ensure the ensemble unity" (Sandu, 2007a, p.162).

And here we are, at the end of our chapter, where we can sum up, concluding what consciousness consist of: specificity and identity with respect to the past with its roots and popular culture, but also with a creative and innovative approach to the present, in order to create for the future in a manner that is coherent and suitable to stand up for hundreds of years without fading the local specificity and identity and to build upon a consciousness that is a cornerstone for the entire nation. But, as Lucian Boia argued, "the Romanian nation, like any other, is less unitary in reality than in the imaginary. It was assembled from distinct parts, welded together by the ideology and centralizing force of the unitary national state (a process made easier by the fact that Romanians were in the majority in all its regions), but even today we cannot talk of complete homogenization (which would, in any case, be impossible)" (Boia, 2001, p12). This is in fact the ingredient that makes the nation, any other nation not only Romanian one, a triumph of cultures and spirits into a fluid synthesis of origins, for continuity and unity through ages.

At the level of culture, however, through the perspective given by what was mentioned in this chapter, there appears the issue of its durability and of the way in which, by cultural creation, man manages to consolidate a certain consciousness towards the land he lives on, providing value and meaning to every process of life and routine. In this context, Lucian Blaga raised the issue of the differences between the minor culture and the major culture and which of these has a greater significance. Hence, the quotation bellow is meant to end this chapter destined to define the concept of culture in terms of values and consciousness, leaving room, in what it is to follow, to a somewhat more thorough study regarding the issues of consciousness related to the local and regional level. Consequently, "each type of culture has its own qualities and deficiencies, its advantages and disadvantages. First and foremost, their chances of durability are dramatically different and this is due to their very constitution. We find it fairly certain that a minor culture, born out of permanent improvisation and gurgling spontaneity, but also out of a total lack of the feeling o perenniality, would be able to last in its statics for several thousands of years; whereas a major culture, born precisely out of its thirst to defeat and outrun the visible space and time, is much more exposed, by its dynamics, to catastrophes and disappearance. (...) The minor culture especially keeps man much closed to nature. The major culture drives him away and alienates him from the natural rules" (Blaga, 2011, pp. 346 – 347).

But, as may Romanian philosophers have shown in their works, among which Emil Cioran, Lucian Blaga and Constantin Noica, Romania has a "dissatisfied consciousness" (Noica, 2010, p. 34) and this is probably why there is a perpetual quest for identity and rebirth of specificity. This approach is all the more significant in the contemporary context of events, as the needs to (re)define our qualities and to build up a strong and healthy consciousness represent the essential ingredients in an [almost] lost battle against multiculturalism. And what better way to end this brief intellectual presentation, but using Constantin Noica's words in his statement according to which, "Dissatisfied, this is the way we feel in general. (…) Dissatisfied with what we know we could have been and we aren't yet" (Noica, 2010, p. 34).

7.3 Birth and Evolution of the concept of Cultural Landscape

In order to study the concept of cultural landscape, we have to understand culture as a value creation process, which through the human activity as led to shaping and changing the exterior space. Thus, it is important to remind what Alexandru Tănase mentioned in his studies of culture philosophy, namely that "no explicative point of view on the culture can avoid the reference to two fundamental realities to which it is correlated: nature and society" (Tănase, 1968, p. 31). Thus, we have a clear mention of the place that culture and nature have with the society, the three of them being connected and influenced reciprocally and the change of one leads to the change or adaptation of the others. The culture, associated to the term "landscape", makes reference to a field specific to the human being, namely what has been modified by human intervention or by what we call cultural process. Thus, the culture involves a creative impulse, the human spirit being involved in a mental process. However, this mental side was ignored for a long time by geographers and theoreticians of culture in the cultural landscape study, being introduced relatively late, as a critical reaction to the first interpretations brought to the concept that nowadays we consider as a fundamental one.

Thus, Don Mitchell presents the history of the appearance of the culture study in geography because, as he says, "it is essential that we situate the history of <<culture>> in cultural geography within a social context – a context composed of changes both within the discipline and in the <<outside>> world" (Mitchell, 2000, p. 16). In the vision of Mitchell, it all started with the wish to understand the superiority of the Western culture, as it was considered in the beginning in Germany, then in Great Britain and United States. The author supports that the roots of the Anglo-Saxon cultural geography are found in the theories of the 19th century appeared in Germany, and in the first half of the 20th century the United States imposed in this field, the parent of the American school of cultural geography being considered Carl Ortwin Sauer (1889-1975).

The main directions that the development of the cultural geography followed were contradictory being considered firstly the environmental determinism which intended to connect the social behaviour of the individual with the components of the physical environment and secondly the studies on folklore which were trying to explain the cultural roots of the contemporaneous civilization. This second aspect supposes the existence of a cultural relativism which came to the front especially in anthropology but appeared predominantly in the works of Sauer (Mitchell, 2000).

7.3.1 Role of Environmental Determinism in the birth of the concept

We will hereinafter also remind about the environmental theory which is considered to have influenced the theoretical and conceptual approaches of Carl Ortwin Sauer (Schein, 2004, pp. 11-24). Thus, the environmental theory is part of material conditions of culture as presented by Tudor Vianu (1982). This theory is based on the fact that the environment was considered a determining factor in the way in which various cultures related and were found in different geographical areas. At the same time, he referred to the way in which social behaviour of individuals was influenced by the environment. The adepts of the theory supported the fact that the environment determinants as well as the climate and various other geographical factors were the only one responsible for the development of human culture and social conditions (Czepczynski, 2008, p. 17). But Tudor Vianu reminded that besides them and the cosmic conditioning of culture, making reference to documents from the 17th century of Fontenelle and Montesquieu (Vianu, 1982, pp. 230-233).

However, in the theory of modern geography, the first influences of environmental determinism appear in the works of the German geographer Friedrich Ratzel (Czepczynski, 2008, p. 17) at the end of the 19th century in which it launches the concepts organic state and Lebensraum (living space). At the same time, Ratzel "introduced the term 'anthropogeography' (Anthropogeographie), as a field of integration or merge of culture and nature, creating Kulturlandschaft, or cultural landscape" (Czepczynski, 2008, p. 17). His concepts appear after the well-known work of Charles Darwin, Origin of Species, of 1859 and were influenced by the evolution of biology, arguing that the state was organic (in essence it would have life) and that it would have needed to grow in order to exist, thus being a natural connection between the individual and the environment.

In the United States, the theory of environmental determinism appears in the works of the geologist Nathaniel Southgate Shaler, of the geographer William Morris Davis or of Frederick Jackson Turner. But at the beginning of the 20th century, the one that succeeds to change the situation is Ellen Churchill Semple, professor at Clark University in Worcester, Massachusetts. She made popular the theories of Ratzel and tried to prove that the study of individuals and of groups of individuals can be done scientifically. Semple tried to popularize geography as a social science which studies the individual, human differences, social life and culture. Around the 1920's, the environmental determinism collapsed under the burden of its own contradictions and irrelevances. Thus, it was tried to redefine geography and the initiative of Carl Ortwin Sauer of using theories of culture in order to explain social differences also originated in the German intellectualism which was found at the basis of the environmental deterministic theory. Thus, "The Berkley school of cultural geography, established by Sauer, a professor at the University of California at Berkley, flourished as a reaction to the materialistic and functionalist approach to landscape" (Czepczynski, 2008, p. 20).

As regards Sauer's thesis, Don Mitchell (2000) in his paper *Cultural geography. A critical introduction* said: "Sauer's main purpose was to show that environmental determinism had pretty much got it backwards. It wasn't nature that caused culture, but rather culture, working with and on nature, created the contexts of life. Sauer was especially concerned with material aspects of culture, particularly the landscape, which he saw as manifestations of culture's traffic with nature" (Mitchell, 2000, p. 21). Thus, we can assert that the American geographer based its affirmation on the thesis according to which the proof of cultural variety was the very landscape which was considered a manifestation of the culture that created it.

Don Mitchell was also among the ones which suggested a short description of theories which influenced Sauer and in his work he mentioned as follows: "Far less clear is what, for Sauer, actually constitutes <<culture>> beyond some vague sense of historically, geographically derived <<difference>>. Culture seems to him a <<hole>> - simply an unproblematic <<way of life>> of a people.(...) It is <<culture>> rather than <<nature>> that determines social life, for Sauer, but the exact nature of that culture remains a mystery" (Mitchell, 2000, p. 24). In his theories, Carl Ortwin Sauer was also influenced by the theses of the Americal anthropologist Franz Boas as well as by two students of the latter, Alfred Kroeber (1876-1960) and Robert Lowie (1883-1957).

Apart from Carl Ortwin Sauer, the American geographer John Brinckerhoff Jackson has also had an accelerant role in the cultural landscape study over the ocean. At the age of 41, after having acquired a vast experience in the cultural landscape study, Jackson wrote in the first number of the magazine Landscape in 1951: "Wherever we go, whatever the nature of our work, we adorn the face of the earth with a living design which changes and is eventually replaced by that of a future generation. (...) The city is an essential part of this shifting and growing design, but only a part of it. Beyond the last street light, out where the familiar asphalt ends, a whole country waits to be discovered: villages, farmsteads and highways, half-hidden valleys of irrigated gardens, and wide landscapes reaching to the horizon. A rich and beautiful book is always open before us. We have but to learn to read it" (Jackson quoted in Wilson, 2003, p. 9). Jackson did not obtain any academic title but was permanently interested in the landscape study. Ever since the beginning of his publisher career, he was concerned with the cultural landscape and with the adaptation of theories and concepts on architecture, urbanism and landscape architecture, mostly as personal reactions and critiques to the modernism. In his book, Wilson said as regards Jackson: "The potential role of designers in making a more meaningful environment was a consistent theme in his writing, and he was vitally interested in the cultural and social meaning of architectural design" (Wilson, 2003, p. 13). In 1934, he came to Europe where he had the occasion to observe and study the constructions of the modern architectural avant-garde: Le Corbusier, Mies van der Rohe or Walter Gropius, which he classed as "ridiculous, intellectual architecture" (Wilson, 2003, p. 41). In general, his woks that criticized the modern architecture and the International Style were written under the pseudonym H.G. West, Jackson having the gift of writing very differently according to the pseudonym he chose due to the literature studies he had and to his vast editorial experience. Thus, the affirmations of Jackson (West) mention that modernism would have been the adept of the dogma "that the architect knows better than the client" (Wilson, 2003, p. 43).

7.3.2 Structuralism in the Evolution of the concept

At the same time, in the evolution of the concept and of studies of cultural landscape, a major influence was held by the structuralism. In order to understand the structuralism influence, it is essential to understand the term "structure" which was used. Thus, we remind the fact that the structure refers to mental models of concrete structure from the reality. The models are not always obvious and we need to understand some hidden or profound aspects, as Mark Glazer (2010) declared: "The most difficult aspect of structuralism is that these structures are not based on concrete or physical phenomena as they are in biological or other sciences but based on cultural realities"[56].

In this period, the influences from the field of philosophy in the cultural landscape study are strongly coagulated around the thesis according to which knowledge and singular experience of the individual become the basis of cultural ideas and of social actions shared by the community and of the way in which the social experience influences citizen's thinking and life. Thus, the way of looking at the cultural landscape changes, passing from the statute of passive entity to its active role in social, economic and political processes of the daily life. Henry Glassie (2000) in his folklore and vernacular architecture

[56] The article http://www.utpa.edu/faculty/mglazer/Theory/structuralism.htm, accessed on 08.03.2010.

studies refers to the intrinsic role of the opposition between order (control) and chaos in all changes brought to the environment. His argumentation being influenced by the structuralism of the anthropologist Claude Levi-Strauss who had "the conviction that deep binary mental structures exist and are expressed and reinforced in ways of thinking about the world and human places within it" (Wilson, 2003, p. 16). Glassie considered that people, regardless of their past and culture, built being guided by the practices of their ancestors or by the rigours of physical laws but most of them were simulated by their own experience in the built space. Thus, he mentioned architects, who regardless of the education they followed, had memories of the inhabited space, of the seen images, of the lived experiences and of the types of touched materials. All these, in combination with the professional (Glassie, 2000, p. 18)[57] made the architect a promoter of culture.

In his trial of explaining the main theories of post-structuralism in geography, Jonathan Murdoch (Murdoch, 2006, pp. 4-11) considered that it was necessary to explain the term "structuralism" and then to explain post-structuralism. The structuralist method is thus presented to us by a series of important features that aim at the main aspects such as for instance the way in which the culture is seen. It was considered as having a systemic attribute and that it should be analyzed as a complex and complete social form which combines elements necessary to stimulate creativity and action but in a closed environment in which societies and groups belonging to a certain culture are considered individually compared to other societies. The analysts tributary to the structuralism were considered as researchers and observers detached from the analyzed object and their investigations which concerned the society were directed to the involved mechanisms and not to social actors or individuals. Thus, the active role of the reader appears when determining the meaning of a research, being important the relation between the reader and the written text, its meanings being multiple and the simple research of some formal structures not being sufficient.

7.3.3 Post-structuralism in the evolution of the concept

Besides the influences of structuralism, the post-structuralism has also contributed to the shaping of cultural landscape studies in geography and not only. Thus, firstly we have to remember that most of the methodologies used in postmodern social studies are taken from the post-structuralism. Compared to the structuralism whose theories considered the culture as being independent from

---

[57] Original quotation: "When the builder's attention is narrowed by training, whether in the dusty shop of a master carpenter or the sleek classroom of a university, past experience is not obliterated. It endures in the strange caves of the brain and in the old habits of the muscles as they seek smooth routes through the air. Education adds a layer. In precept and admonition, in pedagogical technique, if not in content, the teacher brings cultural values into the process of transmission".

the immediate meaning of existence, the post-structuralism considered culture as an inseparable attribute of the existence. Thus, structuralists studied structures which were at the basis of the cultural product, such as the landscape and the text, and used analytical concepts from linguistics, psychology and anthropology. The post-structuralism is hard to define since it completely denies the definitions from the past. The unique, singular meaning of any studied object is unconceivable as well as his unique existence, using varied theories and perspectives which lead to interpretations most of the time conflicting. Thus, ay way of value communication, even by the language of the cultural landscape, is bearer of an ambiguity that makes its precise, scientific interpretation practically impossible.

In philosophy, the post-structuralism has appeared at the beginning of the 1960's and the most important names that we should mention here are Derrida, Deleuze, Foucault, Lyotard or Kristeva. At the same time, this movement has its roots strongly developed in a more distant (philosophical) history, reminding here the influences of Husserl's or Heidegger's phenomenology, Freud's analysis or Kant's transcendental philosophy. The post-structuralism brings new methods of analysis, investigation and understanding of the surrounding world and the human existence. Thus, James Williams (Williams, 2005, pp. 15-17) provides a thorough description of the main critiques brought to post-structuralism stating at the same time that "in poststructuralism, life is not to be defined solely by science, but by the layers of history and future creations captured in wider senses of language, thought and experience" (Williams, 2005, p. 16).

It is obvious that the cultural landscape may be understood as a cultural concept which supposes the approach of varied types of analyses in order to be interpreted and understood being in itself a code of knowledge.

7.3.4 Paradigm change from the Second Half of the 20th Century

Following the failure of the post-structuralism and of critiques brought to this movement, we can consider the period of 60's and 80's of the 20th century as defining phase in the coagulation and formation of several fundamental concepts and of theories which characterized the continental thinking. We can consider as an important aspect in the cultural landscape study the way in which the new global economic order and the bases of the capitalist consumption society, as well as the industrial and information revolution, have influenced the evolution of human thinking and culture. Thus, a turning point was at the beginning of the 60's of the 20th century and the cultural moment that followed at the beginning of the 80's caused changes in the way of perceiving and studying the cultural landscape and the cultural geography as a science managed to surpass the doctrines of the American school whose parent was Sauer and to adopt new theories in the cultural landscape study. At the same time, Rob Shields assimilates this moment with the appearance of phenomenology in the studies of human geography (Shields, 1988, p. 29) which led to an attitude change towards the importance that the individual's experience has and the transformation was made by the voices of some important geographers such as Yi-Fu Tuan or Anne Buttimer (Shields, 1988, p. 29).

For this purpose, the new cultural geography was going to deal with the study of processes and cultural significances of the individual's[58] (Bell, 1978, p. 49) and community's life at different spatial levels involving besides geographers and specialists from several branches such as theoreticians of culture, anthropologists, sociologists and last but not least architects and urban planners.

The theoreticians of the new cultural geography are based on a concept of a much more complex cultural landscape, as a set of cultures forming the real world whose history we must perceive and understand in correlation to the material surrounding reality. A reality which, with the changes brought the post-modernism, is associated to some new parameters which as Fredric Jameson also underlines, have led to the disappearance of the individual as entity and to the appearance of the community, of this new entity to which urban life reports. Jameson stated that this meant the "death of the subject"[59] (Jameson, 1998, p. 5) or the disappearance of individualism. Thus, at the urban and social level, making a discourse on the cultural landscape means making a discourse on community and the demographic increase and the changes suffered by the cities, as well as their expansion and the economic and social context, led to the creation of cultural contradictions.

Consequently, a series of new concepts and theories developed and from the methodological point of view, the cultural landscape analysis and understanding integrated the aspects related to religion, politics, material and immaterial culture, all as a denial and a critique to the American school of cultural geography initiated at Berkeley by Sauer and that of interpretations subsequently approached by J. B. Jackson. After the 90's, we can notice an increased interest among the researchers compared to the urban and the theoretical models were influenced by the Marxist doctrines so that the concepts of cultural landscape launched by Carl Ortwin Sauer and John Brinckerhoff Jackson were considered "too evasive about systematic forces of political economy in mainstream capi-

---

[58] Original quotation: "The individual was considered unique, with singular aspirations, and life assumed a greater sanctity and preciousness. The enhancement of the single life became a value for its own sake".

[59] Original quotation: "But now we need to introduce a new piece into this puzzle, which may help to explain why classical modernism is a thing of the past and why postmodernism should have taken its place. This new component is what is generally called the 'death of the subject' or, to say it in more conventional language, the end of individualism as such".

talist America and in answering the question of who and what, in fact, create urban and rural environments" (Wilson, 2003, p. 16).

Phil Hubbard stated as regards the vision of Trevor Barnes on the theory evolution in geography and the way in which it influenced the cultural landscape understanding and interpretation ever since the first theories of Sauer: "Trevor Barnes offers a similar definition, suggesting that geographic theory need not provide logical relations, rules of causation or empirically verifiable statements. For him, what is crucial is that a theory expresses a phenomenon through a new vocabulary and syntax which changes the way people interact with it (in terms of how they view it, or study it, or practice it)" (Hubbard, 2006, p. 10).

Consequently, as the concept evolved and was subject to changes brought by the structuralist, post-structuralist, deconstructivist or post-modern doctrines, the authors tried to take distance from the "rurality" to which it was associated in the beginning the operational concept of cultural landscape and used new terms, appropriate to the contemporary philosophical speech, such as the one of space or spatiality and place. The evolution was inevitable and certainly legitimate and the first concepts became victims of the critiques of the generations of culture theoreticians that followed. The term "culture" was always ambiguous, just as the cultural landscape launched for the first time by Sauer. Thus, antebellum theoreticians, among which Sauer and to a certain extent Jackson, saw the culture as a complex independent of the human behaviour but responsible for the latter and since the first part of the 20th century were suggested new theories and ideologies which changed the flow of the landscape study and multiplied the meaning given to culture.

Thus, we reassert the fact that in our vision the operational concept culture represents a system of relations, one of power in conjunction with economic, social and political forces. The same can be said about the landscape, either urban/rural, anthropised or non-anthropised, although we can rarely speak nowadays (unlike Sauer and Jackson or the theoreticians at the end of the 19th century and the beginning of the 20th century) about a virgin landscape, unaltered by the human action (but we have to remember the fact that such landscapes still exist). Thus, culture and landscape are sources of power (and sometimes of domination) and combined the two of them form relations that engender some processes (or their result, being at the same time dynamic and stable) which (re)produce spaces and meanings.

7.4 The Cultural Landscape: definitions and ways of understanding

Trying to place the culture at the centre of study of the geography as a science, Carl Ortwin Sauer has published in 1925 its famous essay entitled The Morphology of Landscape, in which he presents a criticism of the natural determinism. In the above mentioned paper, the author suggests the description of the term "cultural landscape" as: "The cultural landscape is fashioned from the natural landscape by a cultural group. Culture is the agent, the natural area is the medium, the cultural landscape the result" (Sauer quoted in Wilson, 2003, p. 5). This description was quickly assimilated by specialists and used as a definition of the term for a long time. However, we have to remind the fact that the term "cultural landscape" was used for the first time in the academic environment by Otto Schlüter in 1908 (James and Martin, 1981, p. 177.). At that time, the German geographer did not provide a definition for the cultural landscape but only mentioned the two types of landscape which should be studied in the geography as a landscape science. Therefore, the author classified the landscape as consisting of the original landscape (*Urlandschaft*) and of the cultural landscape (*Kulturlandschaft*).

Nevertheless, Michael Jones attributes the first references to the term "cultural landscape" to the German geographer Friedrich Ratzel (Jones, 2003, p. 21.) and admits the diversity of the concept that began to characterize it ever since before the Second World War both in Europe and in the United States of America (Jones, 2003, pp.21-22). The author presents several definitions used in geography for this term but besides this fact, what is important to underline is the fact that he generally establishes two categories of use of this term. Therefore, the first would be the category in which the term is used a binder between the various meanings of the culture and of the society[60] and the second category of use would be the one in which the term is simply used as a replacer of the landscape (Jones, 2003, p. 22). But event this distinction of the ways of using the concept has led to the use by those who were the adepts of using the term according to the first category, only of the simple term of landscape in their papers. Therefore, "it is noteworthy that Rowntree's (1996) presentation of the concept of cultural landscape in American human geography ends by talking more of landscape without the adjective than of cultural landscape, while Cosgrove (1998), throughout a chapter entitled Cultural landscapes in a book on European geography, consistently talks of landscape rather than cultural landscape" (Jones, 2003, p. 23). We must emphasize the fact that it is not an error what Michael Jones observed in relation to the approaches of the two geographers, one American and the other one British. What is important to remember

[60] The author mentions here geographers such as Lester B. Rowntree or Denis Cosgrove as adepts of this way of using the term.

from this affirmation is that the term "cultural landscape" and the term "landscape", as we will see in the following chapters, have become almost synonymous, the specialty academic literature acknowledging that the whole landscape has a cultural significance. Therefore, most of the times (but not always), by talking about landscapes we can understand the reference to cultural landscape, without making this distinction[61].

It thus results the importance of determining the significance of the term "cultural landscape" which was used as a meta-word[62]. As a consequence, it is necessary a legitimate question: why do we use the term "cultural landscape"? From our point of view, in the construction of the phrase have been used two words which are among the most ambiguous in the common vocabulary. Both the term "landscape" and the term "culture" are word having a great number of definitions and meanings. As a consequence, their combination given by the phase "cultural landscape" could not be easier to understand or characterized by a single definition. Therefore, this is where can be found the need of knowledge of an extended theory of the concept and of its study.

But before presenting the definitions and the ways in which we make reference to the concept "cultural landscape", we must add a detail that we consider quite important. For this purpose, we must admit the fact that Michael Jones designates a series of concepts used in geography as being "chaotic concepts" (Jones, 2003, p. 24). However, these concepts are not specific only to the geographical study. They are at the bases of many theories in architecture and urbanism because the "geographer seems to be attracted to complex concepts expressed through terms such as landscape, cultural landscape, nature, environment, region, place and space. These are concepts that can be defined in many ways, and they lead to endless discussions" (Jones, 2003, p. 24). And for the study of these concepts, Michael Jones presents two methods of approach. The first approach would suppose elaborating clear definitions for each of the terms, definitions that should help to a better understanding and to the reduction of their ambiguity. The second approach would suppose finding clear meanings, not as rigid as the definitions, but that could lead to a global under-

---

[61] On this occasion, we have the hope that we will manage to answer to all criticism addressed to our research and to the way in which we used the terms "landscape" and "cultural landscape". We think it is a correct argument emphasising the fact that including the personalities who contributed most of all to the development of the concept "cultural landscape" have ended up by using it in association with the concept "landscape", even to the detriment of the former. It is exactly the difficulty of defining the two terms and their ambiguity, as we mentioned several times, that represents the two features making their study an interesting and captivating one. Last but not least, an open, unconfined and free study which, once it has been accomplished, has the power to open minds and redefine ways of thinking on everything that surrounds us. Thus, we hope that our way of approaching the two terms in this paper will be understood.

[62] The term "Meta" represents an element of composition meaning a change. As it is used in the case of metaphysics – meaning here what followed after the physics – in our case, we underline the idea of transformation governing the word "cultural landscape". This term is more than a word, it determines more than a simple meaning. It is a motor of transformation.

standing on different ways to use and define concepts (Jones, 2003, pp.24-25). Therefore, "the argument here is that a systematic presentation of a term's different meanings and usages can contribute to better understanding, communication and dialogue by making the users aware of their differences in their understanding of concepts" (Jones, 2003, p.25).

At the same time, Michael Jones builds the script according to which the concepts are part of a more general discourse on the method of understanding a phenomenon (Jones, 2003, p.25) and "in a pluralistic understanding, concepts and terms can be expected to change in meaning over time" (Jones, 2003, p.25). As shown by the author, this leads to three general ways of approaching the concept "cultural landscape", a narrative-descriptive one, a historical-narrative one and a personal-narrative one (Jones, 2003, pp.26-27). Jones uses the term "narrative" for these types of approaches by explaining the fact that within a general discourse of the cultural landscape there is always involved a narration, a sort of complex story which must be narrated by means of the three ways of approach. Therefore, "a discourse can be broadly understood as any discussion or exchange of ideas, expressed through conversation and dialogue, talks and lectures, and/or writings that treat a subject systematically" (Jones, 2003, p.25.) or quoting Barnes and Duncan discourses are "frameworks that embrace particular combinations of narratives, concepts, ideologies and signifying practices, each relevant to a particular realm of social action" (Barnes and Duncan quoted in Jones, 2003, p.26). These discourses and narrations specific in the definition and the study of the concept "cultural landscape" have finality only if it is considered that "an important point is the role of the researcher in judging what facts are important to select and emphasize in presenting a narrative" (Jones, 2003, p. 28). This is where the solidity of understanding the concept is found and, as a result, the exact way in which we will approach such a study.

In conclusion, we must assert the fact that in this thesis we will be tributary to both above mentioned ways of approach determined by Michael Jones on the concept. Thus, we will have two subchapters for the presentation of definitions and of ways of understanding the concept, one in which we present the approached that we named academic and a second one in which we present a series of definitions from different official documents. But, in order to clarify this aspect, we mention the fact that we consider the second approach more appropriate for the studies of the cultural landscape within our profession, but wishing to present a coherent approach of the two methods, we will also offer an operational definition for the concept "cultural landscape" at the end of this chapter.

7.4.1 Academic approaches in defining the concept

For the purpose of an extended knowledge of the concept, the geographer David Jeffrey Blankenship (2011) builds the script according to which he argues "that the cultural landscape idea was a refuge for the diminished landscape idea (as theorized by Cosgrove) and at the height of 20th century modernism began to transform once again into a new landscape idea that is currently informing contemporary theory and practice in a number of fields" (Blankenship, 2011, p. 16). The cultural landscape has thus become a metaphor and "there are potentially as many ways of seeing as there are eyes to see" (Jackson, 2003, p. 177).

At the same time, Tim Ingold (Ingold, 1993, pp. 24-174) disputes the notion "(cultural) landscape" as a datum of cultural and symbolic nature, thus separating the two worlds, the mental and the material ones. Instead, he suggests a temporality of the landscape that should incorporate processes of the social life and that should prefigure it as part of a dynamic process, of continuous exploitation. "Thus the landscape is always in the nature of <<work in progress>>" (Ingold, 1993, p. 162). For the latter, the notion "cultural landscape" comes from the notion "taskscape" – a series of cyclical, interconnected processes, constituents of the landscape and event structuring parts of the latter. The theory suggested by Ingold and the used term have managed to be accepted as an important way of meditation on the landscape. Another author, Christopher Tilley, accepting Ingold's theories, considers that the binary method (nature/culture) of reflection on the landscape does not contribute to a better understanding. Tilley (Tilley, 1994, p. 24) supports a theory of recognizing the cultural constituent of the environment constitution as if both of them (nature and culture) were the "prologue" and the "epilogue" at the same time. Therefore, both for Tilley and for Ingold, the landscape is permanently incomplete. For that matter, the former develops a sophisticated theory of understanding the landscape as "spirit of the place [...] individual and social times of the memory" (Black, 2003, p. 8).

Figure 16. Mirror of the society - Vienna, panorama. Copyright: Alexandru Calcatinge.

The cultural landscape is thus the mirror of the society reflecting all the attributes, features and characteristics of our existence by means of spaces created based on the knowledge acquired in time. Moreover, we can mention what Tim Cresswell asserted related to the studied concept. "The term is somewhat oxymoronic. Landscape, on the one hand, appears to encapsulate the notion of fixity – of a text already written – of the production of meaning and the creation of dominating power. Landscape is solid. Practice, on the other hand, is about fluidity, flow and repetition. It is about the negotiation between continuity and change. Practice has been seen by social and cultural theorists as an antidote to the representational – as an unexamined component of the everyday" (Cresswell, 2008, p. 4). Compared to it, the term "cultural" which denotes par excellence flexibility, fluidity and dynamics, may seem associated in a wrong or simply accidental, facile and innocent manner. The same author explains the fact that "While there is no doubt that Sauer had a broader agenda than the straightforward mapping and description of different landscape morphologies, it is equally clear that the majority of work by his followers has been centered on just such projects. To Sauer landscape was geography's naively given phenomenon. We often forget that Sauer looked to phenomenology to construct his view of the discipline" (Cresswell, 2008, p.5). This is how, by what we asserted above, we hereby prefigure the exact purpose and importance of understanding the cultural landscape by means of the concept theory.

The geographer Michael Jones (Jones, 2003, pp.29-32) presents the way in which he made contact for the first time with the term "cultural landscape" and the way in which its different meanings have been related during his studies. Therefore, the author presents different ways in which the concept "cultural landscape" is understood in Norway ever since its first appearance in 1934. He essentially distinguished seven different ways of understanding the concept in the specialty literature from the above mentioned country: "1. as an areal category (contrasting with natural landscape and sometimes urban landscape); 2. as a chronological stage of development; 3. as the human components in any landscape; 4. as countryside; 5. as heritage; 6. as scenery with esthetical qualities; 7. as elements in the landscape with meaning for human activity" (Jones, 2003, pp.30-31). At the same time, the author classifies the seven types of understanding the cultural landscape into three great categories: "as landscape modified or influenced by human activity (the first three usages above); as valued features of the human landscape that are threatened by change or disappearance (the fourth to sixth usages above); as elements in the landscape with meaning for a human group in a given cultural or socio-economic context (the seventh usage above)" (Jones, 2003, p.31). In order to exemplify, we consider that it is useful to further remind the way in which the author narrates his experience with the architects and the landscape architects in Norway of the years '70 in the 20th century. Ever since that moment, it was noticed the fact that the professionals of the two disciplines understood the cultural landscape in different ways than in geography, thus "often the cultural landscape meant simply the agricultural or rural landscape, lying between the natural landscape of the mountains on one side and the urban landscape on the other" (Jones, 2003, p. 30). But this changed, as mentions the author and the evolutions from the ecology, history or agricultural sciences have determined a different understanding of the concept "cultural landscape" and "the cultural landscape was understood as the combined product of nature and human impact" (Jones, 2003, p. 30).

Having thus detailed the general framework of the study of the concept "cultural landscape" from the perspective of the geographer Michael Jones, we only have to mention his personal position in relation to the way in which he understands the concept. The author asserts the fact that he sees the cultural landscape as having two dimensions, a physical one and a mental one (or cognitive, as it appears in his descriptions), both of them being equally important because "in practice, nearly all landscapes show signs of the work both of natural forces and of human agency and society, and these two sets of influence are often difficult to disentangle from one another. At the same time all landscapes are cognitive, with varying meanings for different groups of people" (Jones, 2003, p. 32). But what is the most important thing is may the fact that Michael Jones considers the cultural landscape as being more a perspective than a 'thing'" (Jones, 2003, p.32)[63]. A perspective from a certain point of view, the one of the researcher. We will detail this throughout our thesis and it will represent a central element of the research since we think, as we also mentioned in the beginning, that a good practice is always based on a very well built theory and, as regards the study of the cultural landscape, this consists exactly of the way in which we know the concept.

Another specific approach of geography as regards the term "cultural landscape" is tributary to the influences of the French school: "the contemporary cultural landscape approach in geography – sometimes called the regional studies approach – was initiated in France. It was later adopted by several American geographers, who argued that each region has its own distinct landscape that results from a unique combination of social relationships and physical processes" (Healy, 2011, p.5).

---

[63] For a better understanding of what is supported by the author, we will offer here the complete fragments from its text: "cultural landscape thus has both a physical dimension – traces of human activity in the landscape as humans have influenced and modified 'nature' through time; and a cognitive dimension – the cultural meanings that humans attach to their physical surroundings, both natural components and human components. (...) The way in which the physical surroundings, whether of natural or human provenance, are structured and shaped in people's minds begins with the naming of landscape features and places, itself a very culture-specific practice. Hence cultural landscape is more a perspective than a 'thing'".

Figure 17. Agricultural cultural landscape at the border between Austria and Czech Republic. Copyright: Alexandru Calcatinge.

An approach of the term different from the above mentioned ones refers to the perception of the cultural landscape as agricultural landscape. Therefore, in the paper edited by Martin Dieterich and Jan van der Straaten (2004) entitled Cultural landscapes and land use, it is mentioned that "the term cultural landscape has been defined many times by international, national and regional institutions. Many definitions usually create vagueness in the meaning of a term. This book certainly is not about (re)defining the term cultural landscapes. We perceive cultural landscapes as landscapes that are cultivated and/or clearly reflect a history of cultivation" (Dieterich, van der Straaten, 2004, p. Xvi). Remaining in the same field of study, we will mention another paper regarding the agricultural cultural landscape. Thus, in the foreword of the book *The Cultural Landscape. Past, Present and Future*, Krut Faegri stated that "the concept of cultivated landscape was born as a landscape in apparent equilibrium, where man's influence was only one of several equally strong influences" (Faegri, 1999, p. 1). The author continues, reasserting the fact that too little of the existing landscape is unmodified by the man which can also be noticed in case of the cultivated landscape and therefore "a virgin landscape was a fiction. With some small and doubtful exceptions all vegetation types were created or modified by man" (Faegri, 1999, p. 1). This something we asserted several times throughout our paper even though not related to the cultivated landscape but to the fact that there is no virgin landscape left. Therefore, in case of biological studies, Krut Faegri mentioned as a conclusion that the term "cultural landscape" was assimilated in time and more or less naturally since the Second World War (Faegri, 1999, p. 2). It is revealed to us that the term appears however defined in a Swedish dictionary ever since 1939 and is recognized in 1960 for the first time in Norway as an official term (Faegri, 1999, p. 2).

At the same time, in the field of heritage preservation for instance, the understanding of the concept is also different from the above mentioned ones. We will thus refer to the paper edited by Richard Longstreth (2008) entitled Cultural landscapes: balancing nature and heritage in preservation practice. The foreword of the volume describes the way in which the concept is understood and used in the field of the heritage preservation by international theoreticians and practicians. Ever since the beginning, it is indicated the fact that the term "cultural landscape" is simply used many times by the term "landscape" and "is frequently misconstrued as being synonymous with designed landscape – a garden, park, campus, boulevard system, and the like – or with landscaping – the act of manipulating topography, ground surfaces, and plant material" (Longstreth, 2008, p. 1). It may be added the fact that sometimes the term is applied especially to settlements and the countryside. Thus, "the idea of a cultural landscape is at once simple and complex. Studying cultural landscapes entails natural and man-made components of the environment and the ways in which they have changed over time (...). How these components relate to one another is a central point of the inquiry as well. Those relationships may be harmonious, or they may be disparate, even conflicting (...). The significance of the landscape may be rooted in a single event or in a slow, gradual process" (Longstreth, 2008, p. 1).

We must also include in our description the association between the cultural landscape and the tourism, an association often found in the contemporary specialty papers. And not only. We can also find the association in academic, social and political environments. The definition of cultural landscape being more and more assimilated as a resource, it was practically inevitably that this fact should not lead to its use for tourism purposes. "The reality, however, is that tourism is a cultural process as much as it is a form of economic development, and the destination of the tourist and the inhabited landscape of local culture are now inseparable to a greater degree" (Ringer, 2003, p. 2). Thus, as Mariusz Czepczynski asserted that "cultural landscapes can be seen as being among the main composers as well as vivacious transmitters of culture. (...) Cultural landscape is one of the main representing languages of modern society, which signify the spiritual dimension of the investors, architects and users" (Czepczynski, 2008, p. 28). At the same time, "the surrounding is central to understanding the landscape in that it embodies economic, social and cultural processes" (Czepczynski, 2008, p. 28).

Considering the elements presented until now, we can notice that there are many definitions given to the concept studied in our thesis. Therefore, Mariusz Czepczynski states the fact that "it is hardly possible to present a holistic, comprehensive contemporary definition of cultural landscape, since both terms, culture and landscape, carry so many meanings, explanations and understandings, and the melange of cultures and landscapes creates dozens of possible, used and interesting definitions" (Czepczynski, 2008, p. 27). This argument of the Polish geographer appears when he previously accepted the fact that "Cultural landscape was generally understood and defined as the human-modified environment, including fields, houses, churches, highways, planted forests and mines, as well as weeds and pollution" (Czepczynski, 2008, p. 20). Thus, we must accept, as we will show several times in this paper, that the ambiguity of the term is the one granting it vitality in its more and more frequent usage in the academic, social and political environments. Has this meta-word become a symbol? Has the (cultural) landscape become and existed itself as a symbol? For this purpose, as well as in order to complete the argument related to the definition of the term "cultural landscape", we must underline the fact that (cultural) landscape is a symbol exactly due to its significance full of different and various meanings. It is sometimes a dynamic concept but also static, as we will mention throughout our paper. It is a process and a product at the same time. It is a mental construction. Therefore, its understanding as symbol is due both to the way in which it is defined and to the way in which it is operated.

Donald Meinig said that "we regard all landscapes as symbolic, as expressions of cultural values, social behavior, and individual actions worked upon particular localities over a span of time. Every landscape is an accumulation, and its study may be undertaken as formal history, methodically defining the making of the landscape from the past to the present. (...) And every landscape is a code, and its study may be undertaken as a deciphering of meaning, of the cultural and social significance of ordinary but diagnostic features" (Meinig quoted in Backhaus, 2009, p. 11). In the contemplation of this comprehensive and at the same time permissive description, Gary Backhaus thus remind the fact that (geographers) "do offer intensive definitions, but there are working definitions, offered in case study testing grounds, which lack in the elaboration necessary for theoretical demands of rigor" (Backhaus, 2009, p. 7). Nevertheless, the working definitions have helped every time in the study of the (cultural) landscape. The way of relating with the concept, even in the lack of a single, clear and holistic definition, was based on different meanings which were given to it throughout time. But we can add the fact that this did not lead to a clarification as regards the term "cultural landscape" but only to a deepening of its ambiguity which led to the adoption of several working definitions according to the discipline dealing with its study. Or even the use of several definitions within the same discipline.

Figure 18. Mirror of the society - Bucharest, Arcul de Triumf (Triumphal Arch). Copyright: Alexandru Calcatinge.

Figure 19. Mirror of the society - town of Izola, Slovenia. Copyright: Alexandru Calcatinge.

One of the approaches we find in the study specific to the urbanism and landscape school from our country is given by Mr. Professor Cătălin Sârbu, who stated the fact that "if we approach the cultural landscape as the complex expression of a way of life expressing synthetically the report between the natural and the anthropic resources, we will identify more easily the elements of a dynamics sometimes subtle, other times more abrupt but which represents an expression of the becoming of a human community together with the territory on which it lives" (Sârbu, 2011, p. 9). The same author states the fact that "the history, the culture and the territory combine in a process of development in which the cultural landscape may be considered as one of the structures of the territory. (...) If we understand the cultural landscape in the sense of the elements ex-posed so far, we realize that it is about a live territorial reality which exists first of all by the execution of activities of the population found in that area (...) the cultural landscape becoming a central element of the local development strategy" (Sârbu, 2011, p. 11). As Mariusz Czepczynski also asserted the fact that it is difficult to give a contemporary holistic definition for the term "cultural landscape", Cătălin Sârbu stated the fact that "the most important contemporary phenomenon consists of the loss of the holistic understanding of the landscape by its inhabitants" (Sârbu, 2011, p. 13).

Figure 20. Mirror of the society: Hamburg. Copyright: Alexandru Calcatinge.

For the idea of understanding the concept "cultural landscape", we must remind an approach common to many colleagues that we consider insufficiently argued and certainly perfectible. We meet quite often among our students the reference to three types of landscape: anthropic, natural and cultural, as distinct and different categories. However, we have to state the fact that in our acceptance the cultural landscape is a result of the other two types of landscape, anthropic and natural. For a better exemplification, we will offer hereinafter the description of the types of landscape made by Mr. Professor Wilfried Schreiber in Cluj-

Napoca: "The landscape is a dual, objective and subjective reflection of some combinations of elements. (...) Initially, the landscape was a natural one, namely unmodified by the man. (...) at the present moment, natural landscapes can only be found in several high mountain regions and in the Polar Regions. On the continents populated densely and for a long time the landscape is an anthropized or anthropic one. An anthropized landscape refers to an initially natural landscape, modified by the man, which still keeps however elements of the natural landscape (...). A completely modified landscape or the one built by the man (for instance, a city, a railway system or an industrial landscape) is considered to be an anthropic landscape. Both types of landscapes, the anthropized and the anthropic ones, are known as cultural landscapes" (Schreiber and Stoica, 2008, p. 7).

In conclusion, we can thus clearly distinguish the fact that the approaches of the term and concept "cultural landscape" will always be various, just as its definitions. Its uses in the academic, university or political environment will also be various. Nevertheless, from our point of view there is a common element that connects all the types of theoretical approaches and materializations: the need for a conceptual clarification of the term "cultural landscape". We think this is the purpose of its frequent use in different studies and researches specific to some related fields among which we can mention architecture and urbanism, together with geography, anthropology, philosophy, cultural studies etc. Certainly, we have not presented all the definitions given to the term "cultural landscape" because we think that they all will not lead to a better understanding of the concept. As Michael O'Flaherty[64], among others, stated: "It's too bad the term <<cultural landscape>> needs to be defined and defended. As an anthropologist (...) I take it for granted that all land(scape)s, (...) are innately cultural; human beings are innately cultural, (...) and wherever people have travelled they have entered the land into their imagination and, insofar as this has been a collective endeavour, into their culture" (O'Flaherty, 2010, p. 1). His conclusion is also maybe the one that should interest us as architects and urban planners. Here is what states the Canadian anthropologist: "Since all people have a (cultural) relationship to the land and its other inhabitants, the term <<cultural landscape>> is redundant. Our goal should not be to defend the simple fact that landscapes are cultural but to find ways to describe the myriad of relations and their interaction with one another" (O'Flaherty, 2010, p. 1). It is this matter of identifying the relations between cultural landscapes and their components that is important in a study of the cultural landscape so that it has validity and utility, beyond the simple identification of categories. However, this can be realized from our point of view only by a thorough knowledge of the theory revolving around the concept and without which the discovery and the understanding of connections and relations between the components of the

[64] Michael O'Flaherty has a Doctorate degree in anthropology and is a Professor at the University of Manitoba, Winnipeg, Canada.

cultural landscape cannot be complete. But as one can notice from this chapter, the multitude of definitions given to the cultural landscape can disarm any individual interested in the study of the concept. We must supersede these phases and realize the fact that a good operation of the concept "cultural concept" does not result only from the understanding of its definition, but from the understanding of all theories, definitions and paradigms which led to the enrichment of the operational concept. Therefore, for any study of the cultural landscape which is based on these information categories, each of us will be able, according to the discipline of study, to outline an operational definition which should underline the purpose of the analysis in the most concrete way.

7.4.2 Definitions of the concept in some Official Documents

Besides these short descriptions of the way in which is understood the concept "cultural landscape", in the specialty literature there are many definitions or descriptions of the latter operated by specialists in different fields. We will certainly not remind and present all of them but we will offer a description of those that we consider important for the support of the arguments in our study.

First of all, we will refer to the European Landscape Convention. Although the expression of cultural landscape is not defined, there still is a series of descriptions in which the landscape and the culture are associated. Therefore, at the exact beginning of the document, in the preamble, the convention elaboration is justified by reminding some guidelines regarding the landscape, such as among others: "(...) that the landscape is an important part of the quality of life for people everywhere: in urban areas and in the countryside, in degraded areas as well as in areas of high quality, in areas recognized as being of outstanding beauty as well as everyday areas, noting that developments in (...) techniques and in regional planning, town planning, transport, infrastructure, tourism and recreation and, at a more general level, changes in the world economy are in many cases accelerating the transformation of landscapes, (...) that the quality and diversity of European landscapes constitute a common resource"[65]. Everything mentioned above has led to the definition of some terms in the first article of the document. Among these, is defined the term "landscape" as "an area, as perceived by people, whose character is the result of the action and interaction of natural and/or human factors"[66]. Thus, a cultural meaning is given to the landscape by adding the terms "natural" and "human" and by underlining the fact that the two factors are results of the some interactions and actions. This could lead us to what Christopher Tilley said and that we reminded earlier in

---

[65] European Landscape Convention – fragment taken from the website: http://conventions.coe.int/Treaty/en/Treaties/Html/176.htm.

[66] European Landscape Convention – fragment taken from the website: http://conventions.coe.int/Treaty/en/Treaties/Html/176.htm.

this chapter. Nature and culture are a sort of prologue and epilogue of a specific work based on interactions which determine the character of a landscape. Even the arguments of Blankenship according to which the cultural landscape is only a deliverance of the usual term "landscape" thus receives an official support, not only the academic already recognized one.

We remain in the field of official definitions and we remind the one given by UNESCO to the term "cultural landscape". Thus, the guide for the registration in the UNESCO World Heritage List, page 88 art. 6, defines the cultural landscape as a collection of "cultural properties and represent the <<combined works of nature and of man>> (...). They are illustrative of the evolution of human society and settlement over time, under the influence of the physical constraints and/or opportunities presented by their natural environment and of successive social, economic and cultural factors, both external and internal" (UNESCO OP. GUIDE, 2011, p. 88). A completion on the understanding of the term is provided in art. 8, specifying the fact that "the term <<cultural landscape>> embraces a diversity of manifestations of the interaction between humankind and its natural environment" (UNESCO OP. GUIDE, 2011, p. 88). At the same time, these cultural landscapes are classified into three categories: (1) landscape designed and created intentionally by man; (2) organically evolved landscape; (3) associative cultural landscapes (UNESCO OP. GUIDE, 2011, p. 88).

We will hereinafter mention an Australian document regarding the cultural landscape management. This reminds several definitions or descriptions of the concept contributing to outlining a valid working definition. Therefore, the first definition mentioned refers to the way in which the term "is applied to those parts of the land surface which have been significantly modified by human activity, to distinguish them from natural or wilderness landscapes, which have little or no apparent evidence of human intervention" (Lennon and Mathews, 1996, p. 3). In the immediately following paragraph, another definition of the cultural landscape is given to the cultural landscape, as being "rural and urban settings (spaces) that people have settled or altered through time. They include cultural and natural elements of the ordinary, familiar, everyday landscapes" (Lennon and Mathews, 1996, p. 3). Thus, the cultural landscape is considered as a mosaic (Lennon and Mathews, 1996, p. 3) consisting of three important elements, such as: (1) natural elements, (2) physical components resulting from human activity and (3) patterns created in the landscape over time. As a consequence, these definitions together with other descriptions made the authors to elaborate a working definition that they used in the above mentioned document. Here is this definition: "A cultural landscape is a physical area with natural features and elements modified by human activity resulting in patterns of evidence layered in the landscape, which give a place its particular character, reflecting human relationships with and attachment to that landscape" (Lennon and Mathews, 1996, p. 3).

7.4.3 Definition of the concept of Urban Cultural Landscape

This chapter was the basis for the publishing the start-up volume *Visions of the real. An architect's approach on cultural landscape studies* in 2011 in Vienna. In the above mentioned book we have approached in a more detailed manner the script according to which the cultural landscape is a vision of the real. We will hereinafter present the main concepts which are at the origin of this approach, but we have to mention the fact that a series of ideas will not be found in this thesis.

Over time, within the urban development processes, there has been identified a series of defining key moments which led to space evolution and shaping we live in nowadays. Currently, under the circumstances of two major forces governing the world, the globalization and the so called new economy, cities and regions have found a new way to coexist and survive in which the technology, the human capital, the creativity and the innovative capacity are decisive elements. We named this moment the cultural moment in which old paradigms of social development made room for a postmodern dimension, the cultural one, essential for the understanding of an environment under continuous transformation.

The postmodernism has criticized a series of concepts of the modernity and, at the same time, has launched new notions and terms in order to define the surrounding realities. They have been easily assimilated in discourses and have become usual notions whose meanings are extremely complex. Under the circumstance, the concept "cultural landscape", a modern(ist) one par excellence, has resisted to constant criticism and has even managed to develop based on it and on the creativity of those involved in its study[67] (Bell, 2001, pp. 297-306).

The contemporary city is characterized by permanent transformations at the level of all elements forming the urban structure and the social life. All these transformations appear in relation to the closeness, to determined events and to the memory of past experiences. Thus, Sharon Zukin offers a steady criticism of contemporary cities due to the fact that "they represent the basest instincts of human society" (Zukin, 2000, p. 132), while for Hannah Arendt "the clear sign that society represents the public organization of the very process of life is maybe the fact that in a relatively short time the new social domain transformed all modern communities in societies of workers and employees" (Ar-

---

[67] Daniel Bell stated: "the modernism was held responsible with no doubt for the great influence of the creativity in the occidental culture. (...) Most of these experiments were triggered by the creative tension of the culture, found in the position of adversary against the bourgeois social structure. However, there was a price for all these. A loss was the lack of coherence in culture, especially in the spread of an antinomic attitude towards the moral norms or even towards the idea itself of cultural judgement".

endt, 2007, p. 43). Therefore, the urban landscape forms what Robert Ezra Park (Ezra Park, 1926, pp.1-14) named a *moral order* or Sharon Zukin a *landscape of power*[68]. The order mentioned by Park was an apparent one which appeared paradoxically due to the disorder. It is a different order, due to the disappearance of a determined urban order which, in order to be creative, had to be destroyed and left behind, the city representing, like in the visions of Engels and Marx, only a particular aspect of a more general problem with a strong sociologic character[69]. According to the two theoreticians, it can only be settled by the appearance of the society with no social classes. A similar problem such as the one presented by Marx and Engels (Marx and Engels, 1970, pp. 39-95) is re-enacted by the moments of social crisis when urban social structures are interrupted and lead to the formation of a new order. The social crisis is therefore an opportunity to progress, the urban landscape being shaped in time and space, the succession of those moments composing the history in which local or regional culture coexists in its own space-time framework, being however affected by the Globalization.

Therefore, in the 20th century, the new undertaking of observation and reflection on the city appeared as a phenomenon exterior to the individuals living in it. The city is seen as a mirror of the modernity which makes us reflect on the condition of the individual as part of culture-bearing and identity-generating social groups. The term "mirror" given to the city refers on the one side to Foucault's heterotopias for which the mirror is a mixed space, simultaneously utopia and heterotopia, "a place without place" but also having a "retroactive space" (Foucault, 2001, pp. 250-260) and on the other side at the ideal citadel as an imaginal city (from the term "*imaginal*" of Henri Corbin[70]), placed beyond space and time.

The urban contemporary life has lead to the appearance of two novel experiences, the one of the abandonment and the one of the loneliness which, in their turn, has caused serious identity problems among the individuals. The postmodern world has made that the uncertainty which led to the citadel construction as a defensive system to become an urban anguish. Therefore, the individ-

---

[68] In Landscapes of Power: From Detroit to Disney World, University of California Press, Berkeley, 1991. In this book, the author applies the concept "creative destruction" in the construction of the script according to which the landscape represents the most important cultural product of contemporaneity.

[69] Marx and Engels were among the first sociologists which discussed about the urban-rural transition and the city evolution in social terms. Their most known ideas on this line have been exposed in the paper The German Ideology.

[70] The use of the term is explained by the French philosopher in the article entitled Mundus Imaginalis, or the Imaginary and the Imaginal (http://hermetic.com/bey/mundus_imaginalis.htm accessed on 20.12.2009) as imperative to be able to accomplish his studies on Arabic and Persian texts, the use of the term "imaginary" being inappropriate due to the meaning and interpretations it can engender, representing something unreal and utopian. Therefore, by the term "imaginal" (or mundus imaginalis) he refers to a certain order of reality with direct implication on the imaginative consciousness.

ual is in a direct relation to an interior aggressor, a relation which is greatly mediated by the derangement of the social order and of inter-human connections but also by the changes of the urban cultural landscape. At the same time, in an era of speed and of financial supremacy, the globalization and the descendant direction of global economy makes us consider the urban culture as not being entirely coordinate by the two factors (money and power) but as a whole determined by the contradictions of the urban life. The society puts in relation the spacio-temporality with the level of technological evolution of the civilization, the cities being thus centres connected with various realities and opening the way to large opportunities. It is generally known the fact that they were ever since their beginning trade or exchange centres, media centres or poles of the global policy, their strategic role gaining a bigger importance in the new era of globalization. As a consequence, we can assert that the urban landscape has always been one of power, of (dis)order and of progress. According to Richard Sennett[71], the contemporary cities are cosmopolite, dense and to some extent dangerous, being spaces of dislocation, destructuring and disorientation. At the same time, the contemporary metropolis is a space of hybrid cultures and heterotopias. These cultural differences are the ones alienating us one from another, causing the feeling of fear and uncertainty in urban spaces. Thus, the reciprocal avoidance and separation tend to become survival strategies in the urban metropolis; in its turn, it changes into a space of censorship and or excessive control where panoptical strategies become possible.

The contemporary urban cultural landscape is a space of identity shaping, active and dynamic. Its perception is complex, of phenomenological nature, being both a cultural process and cultural product. The process of social life is directly integrated into the notion of urban landscape since, according to Eric Hirsch, the concept of landscape is determined by a "is a cultural process that brings together the cultural meaning of landscape with the concrete actuality of everyday life" (Hirsch, 1995, p. 3). The expression "urban cultural landscape" is relatively new and involves knowing and understanding all terms. In many texts, the urban landscape is also named architectural landscape (cityscape or townscape). Thus, one of the important characteristics of architecture is put forward: the creation is never neutral, participating in the identity construction as a component of the cultural landscape by the mental part and the sensibility added to the space meant for habitation and social life.

Pier Carlo Palermo considers that the "the fragmented and sometimes chaotic image of many contemporary urban landscapes probably expresses the uncertain coexistence of a plurality of partial orders, still little understood" (Palermo, 2008, p. 32), referring to the fragmented image of contemporary cities as determined by the coexistence of different cultures and by the relations established between them. Referring thus to the urban landscape we aim in fact the

---

[71] In the online article A Flexible City of Strangers, in Le Monde diplomatique English Edition, February 2001 - http://mondediplo.com/2001/02/16cities accessed on 20.12.2009.

relations. Their dualism may also be seen from the assertion of Russell West-Pavlov regarding the space which "is not a pre-existing container for artefacts and practices, but is constituted by them in a relationship of reciprocal influence and inflection" (West-Pavlov, 2009, p. 24). The relations defining the urban landscape and the coexistence of different cultures inside it are also rigorously defined by Theodor Adorno; for instance, we mention the next paragraph: "if it were possible a sort of psychoanalysis of the culture specific to the present moment, if the absolute domination of the economic did not ridicule any trial of explaining the current circumstances by the spiritual life of their victims and if the analysts themselves did not swear faith to this exact state of things, then a such investigation should demonstrate that the disease of our time is the normality itself" (Adorno, 2007, p. 70). We can notice here a normality perceived differently for each historical period, never the same for the small cities or for each individual. We can actually admit that this is a normality founded on the numerous cultures forming the contemporary society. In this order of ideas, we can ask ourselves what represents the urban cultural landscape for the modern man or, in other words, what represents the cultural landscape from the perspective of the individual which is essentially urban. In order to answer, we must underline the fact that ever since earliest times the architect has represented the world of certainty, of culture and of power by the ability of space structuring. The city being thus just a fragment of the reality. The everyday of the contemporary life thus proves that a purely architectural metaphysical city or a philosophical system are utopian; by his work, the architect proposes a complex experience reuniting time, space, tradition, creativity and senses. Space sensibility is given by the senses with which we perceive it, by the way in which we notice the configuration and the way in which we receive it. Therefore, the space is a system of relations, a territory of experience, in which everything has its place even the normality of existence; understanding the cultural landscape is strictly related to this normality of existence and of space which "juxtaposes or overlaps the <<the traces>>of several previous periods whose primary meanings we miss since every trace is reinterpreted and reintegrated in the contemporary functional context" (Mihali and Copoeru 2002, p. 11). In his paper (Re)reading architectural landscapes, Iain S. Black invokes a interrogative quotation of Olsen from 1986: "what messages were buildings, cities, and other works of art expected to transmit? What meaning did they possess, what ideas did they contain? What can a city, in its capacity as a work of art, accomplish? What can art do, apart from existing in its own right? It can tell a story, or many stories. It can establish a mood. It can reinforce selected virtues, textures, colours and movements. It can stand for, or represent, ideas, qualities, institutions" (Black, 2003, p. 19). In these lines, it is exemplarily underlined the cultural product of the city but at the same time the one of cultural process in which the difference is the engine of the social life of the individual, the one alimenting the dynamics of the urban landscape. A new dimension of the urban experience is shapes, in which the city offers both a public space, of culture, and a family space, a private one. In *Minima Moralia*, Theodor Adorno (2007) presents a pessimistic overview on the city and on the social life from the end of the Second World War: "The destruction of the European cities […] are merely the executors of what the immanent development of techniques long ago decided for houses. These are good only to be thrown away […]. The possibility of dwelling is being annihilated by that of the socialistic society, which […] sets the bourgeois one in motion towards catastrophe"[72] (Adorno, 2007). This reality is shaped from the gloomy perspective of mutations that the technological revolution causes to dwelling as well as on the influence of socialist doctrines over the individual and his social future; at the same time, it is invoked the possibility of denying changes or of accepting them by reconstructing altered cultures: "the thought that […] life could continue on "normally", or indeed that culture could be "reconstructed" – as if the reconstruction of culture alone were not already the negation of such – is idiotic" (Adorno, 2007).

Figure 21. Mirror of the society. The House of Parliament in Bucharest, Romania. Copyright: Alexandru Calcatinge.

We must discuss hereinafter a series of operational concepts introduced by contemporary theoreticians. Thus, Giovanni Maciocco (2008) theorizes two perception levels of the urban cultural landscape for the modern man. The first level is represented by the concept "environmental image" which defines the dynamic environment of the contemporary city, while the second level investigates what could be defined as metropolis counter space (Maciocco, 2008, p. 1). Other authors, such as Iain Robertson (Robertson, 2003, pp. 1-17), discuss

[72] Taken from the website: http://members.efn.org/~dredmond/MM1.html.

about landscape in terms of a cultural product but at the same time also of cultural process, a dichotomy existing between a landscape history and a symbolistic of the latter.

Therefore, the symbol elements of the urban cultural landscape become iconic signs of a local, regional or national identity. Approaching the landscape from the perspective of symbolism presented a great interest for researchers and geographers at the beginning of the '80s in the past century, this analytical direction originating in the idea that according to which the landscape is not nature, but nature transformed by humankind (Sauer, 1963, pp. 315-350). The acknowledgement of this symbolic meaning of the cultural landscape leads to the assumption of a political side of the space and to the determination of its mental and material qualities. Thus, cultural landscapes are nothing else but cultural images often hiding the process by which they appeared (social, political, economic or spiritual) behind some familiar "facades" which seem to be natural.

Figure 22. Mirror of the society. Abandoned industrial site in Mediaş. Copyright: Alexandru Calcatinge.

Regardless of the approached concepts, the urban cultural landscape is determined by tensions and relations put at stake the two fundamental components of space – remoteness and proximity, which are correlated to the moving process and to time. In their turn, the human mind and body are taken by tensions in the process of observation or direct involvement in knowing and reading the landscape. Thus, value systems and ideologies are in direct relation with the permanent adaptation of social life, the individual being in a continuous research of identity in a changing space in which the distinction is the evolution engine. Social events in the large urban landscape are those giving birth to new cultures or subcultures which are different from the existing ones. It is this difference that led to the creation of the contemporary society which made the city a "laboratory" of various configurations, a space of representation, contestation and identification at the same time.

Figure 24. Mirror of the society. Bucharest - House of the Free Press. Copyright: Alexandru Calcatinge.

Each city, as well as each individual, has its own identity, historically determined in time and space. Starting from these data, it can only be a value system the one constantly interacting with the urban system and determining the cultural landscape, the identity representing a way of connecting different evolution phases in the development of the urban system. Taking into consideration the aspects presented in the paper, the contemporary city thus represents a spatial and social cohesion, in a symbolist interpretation. It is an environment which can no longer offer protection, becoming extremely vulnerable and amplifying the panic. Social cohesion is thus compromised due to its cultural dimension itself. Given the aspects mentioned above, the urban cultural landscape can be understood as a collection of relations and tensions; it is ultimately a process determining the contemporary public space as a space of reflection in which the individual can survive without necessarily feeling the manipulation orchestrated by communication flows which standardize the contemporary urban experience.

At the same time, the city is also a mental construction, which reflects ideas and concepts of those who created it over time, being the materialization of the power of visions at different conceptual levels, as well as a support for spaces which are equally imagined and belonging to the imagination both collective and individual. We are facing a paradigm of architecture and urbanism in which subjectivity plays a huge role in decoding the reality and the built-up, the cultural landscape and the ability to understand what is subject to analysis is up to the capacity of each of us to interpret symbols, to decode a message. Thus, we end by interpreting the city as a text, a paragraph describing a mystery, an apparent chaos, a puzzle. The individual is thus subject to the rigours of the contemporary life in which in order to understand the cultural landscape, he needs to acquire the qualities of a real detective.

Figure 25. Mirror of the society. The mountain village of Poiana Sibiului. Copyright: Alexandru Calcatinge.

7.4.4 Definition of the concept of Historical Cultural Landscape

After presenting the ways of understanding the concepts "cultural landscape" and "urban cultural landscape", we have reached the moment when we will also approach a little bit the frequent term "historical cultural landscape". This term is also known as traditional cultural landscape. If we may say so, it is a term which often substitutes the cultural landscape as presented in the previous chapters. Even if it is not always defines as historical cultural landscape, most contemporary works prove the fact that this traditional and historical side, many times mistaken for the rural landscape, is frequent in the current studies.

Figure 26. Image from the village of Gura Râului, Sibiu County, Romania. Copyright: Alexandru Calcatinge.

We will mention a study made by "Babes-Bolyai" University of Cluj-Napoca in partnership with the Ethnographic Museum of Transylvania in Cluj-Napoca. The study refers to historical cultural landscapes in Transylvania and was completed in 2008, the result being published in a book-register entitled Historical Cultural Landscapes (*Peisaje cultural istorice* in Romanian) (Schreiber and Stoica, 2008). Therefore, the authors define the historical cultural landscape as "the one that has been created before the historical phase we are in at present" (Schreiber and Stoica, 2008, p. 7). This is all we mention for our analysis and the construction of the argument suggested in this thesis. As such, the study of the professors from Cluj focused on inventorying the types and elements of historical cultural landscapes. We will briefly present them, emphasizing the fact that many of them will be reminded in the chapter on chronotope, but in different circumstances. But in the same order of ideas we will firstly remind what H.J. Birks asserted, namely the fact that the process of "conservation of the cultural landscape is not solely nature conservation, as it requires preservation of traditional land-use practices, buildings, walls and other components of the landscape mosaic, and of traditional way of life" (Birks, 1999, p. 8). The above mentioned process being described, we have to underline the similarity of this type of approach to the current practices in the cultural landscape studies to which we have been accustomed within our school. Therefore, we consider that we don't only have to make this inventory of traditions and elements of some existing cultural landscapes but rather, as Michael O'Flaherty (O'Flaherty, 2010, p. 1) and Wielfried Schreiber (Schreiber and Stoica, 2008, p. 7) stated, to find and understand the relations between these elements. Therefore, "there is more to cultural landscape ecology than description and documentation of patterns in time and space. There is consideration of underlying causal processes" (Birks, 1999, p. 9). However, before reaching such an interpretation and emphasis of processes and relations between elements, the first step is to mention them. The interpretation of these above mentioned causal relations and processes supposes the knowledge of some theories both in the field of cultural landscape study and of urban, anthropological, ethnographic studies and others. Therefore, one can distinguish the importance of the pluri-disciplinarity of such studies and the need that each specialist should know the study framework specific to each subject.

Therefore, we will hereinafter present elements of historical cultural landscape as mentioned by the team of professors in Cluj-Napoca[73]. Regarding the project we started to describe in the previous paragraph, Jozef Hernik, the project co-ordinator from the main partner in Poland, stated that "the entire project emphasized the importance of interdisciplinary scientific exchange of both students and researchers in order to create networks that would be the forum for the exchange of experiences" (Hernik, 2010, p. 567). We underline this aspect in order to maintain the main idea of cultural landscape studies that we consider to be exactly inter-/trans-/pluri-disciplinarity which should bring together the theories governing the studied concept from all associated subjects and should use them for the common purpose of the research. Thus, Jozef Hernik emphasizes the importance within the cultural landscape study of a good correlation be-tween theory and study methodology. This is realized by presenting some diagrams in which are indicated the main phases in the project elaboration (Hernik, 2010, pp. 568-569). The catalogue of landscape elements and the glossary of terms present a number of 56 elements of historical cultural landscapes[74]. Out of the 56 elements, we would like to emphasize some that we consider more important in our study and which appeared in the book of Mr. Professor Wielfried Schreiber mentioned above. These elements will also help us in the foundation of a more clear understanding of what represents the historical cultural landscape, since a clear definition could not be given so far.

The presented elements have been classified into five different categories, such as: elements with religious significance (Schreiber and Stoica, 2008, pp. 8-11), dwelling elements (Schreiber and Stoica, 2008, pp. 11-14), household elements (Schreiber and Stoica, 2008, pp. 14-20), economics elements (Schreiber and Stoica, 2008, pp. 20-38) and elements related to water management (Schreiber and Stoica, 2008, pp. 38-39). We will only list them, without detailing them.

---

[73] We must underline the fact that this research is part of the Phare National Programme 2004 INTERREG IIIB CADSES and was accomplished by the partnership of several public institutions and universities in Austria, Poland, Germany, Romania and Ukraine. The main partner was the Agricultural University of Cracow and that the two Romanian institutions mentioned previously in the paper participated. The project which is part of this research bears the title Protecting Historical Cultural Landscapes to Strengthen Regional Identities and Local Economies and is part of bigger project entitled Advancing Transnational Co-operation within INTERRG IIIB CADSES.

[74] The source of documents is http://www.cadses.ar.krakow.pl./. Both the register of elements and the catalogue with the glossary of terms can be downloaded from here. These documents are available for the user in the languages of all countries involved in the project: Polish, German, Romanian and Ukrainian

Figure 27. Image of an old church near Roşia Poieni and Bucium, in Apuseni Mountains, Romania. Copyright: Alexandru Calcatinge

The elements with religious significance are divided into:

1. church (with separate components – wooden church, brick or stone church, fortified church);
2. monastery;
3. cemetery;
4. cross and roadside crucifix.

Figure 28. Street of a mountain village near Sibiu. Copyright: Alexandru Calcatinge.

Figure 29. Old village replica in the ASTRA Village Museum in Sibiu. Copyright: Alexandru Calcatinge.

The dwelling elements include:

1. house (with separate components – wooden house, stone and brick house, mixed material house);
2. citadel;
3. castle;
4. seasonal dwellings.

Figure 30. Old historical house replica in the Romanian National Peasent's Museum in Bucharest. Copyright: Alexandru Calcatinge.

The household elements include:

1. hayloft;
2. barn;
3. fence;
4. wall;
5. gate;
6. fountain;
7. bag;
8. stove;
9. small garden in front of the house;
10. garden;
11. cellar.

Figure 31. Typical house from the central area of Transylvania. Copyright: Alexandru Calcatinge.

The economics elements include:

1. Agricultural elements

1.1. field

1.2. agro terrace

1.3. haystack

1.4. grass land

1.5. grazing land

1.6. vineyard

1.7. orchard

1.8. beehive

1.9. market place

2. Forestry elements

2.1. low forests

2.2. bush

2.3. repeated exploitation of trees

3. Industry and handicraft elements

3.1. watermill

3.2. fearnought milling machine

3.3. whirlpool 3.4. distillery

3.5. oil mill machinery

3.6. oil press

3.7. sawmill

3.8. forge

3.9. pottery

Figure 32. Another old house replica in the Peasant's Museum. Copyright: Alexandru Calcatinge.

Figure 33. Wine cellars in Austria. Similar to a typical street in a Saxon village from Transylvania. Copyright: Alexandru Calcatinge.

4. Mining landscape elements
4.1. gallery
4.2. stamp
4.3. refuse dump
4.4. quarry
4.5. gravel plant
5. Transport and traffic elements
5.1. railway station
5.2. railway
5.3. paved/asphalt roadway
5.4. alley
5.5. footway
5.6. ford
5.7. bridge, footbridge

Water management elements:

1. dam
2. channel
3. dike

7.5 Conclusions on knowing the concept of cultural landscape

In the previous subchapters, we enlarged the concept "cultural landscape". We referred to the different definitions and ways of understanding it. We have stopped a little bit on the separate analysis of the concepts "landscape" and "culture" by means of values and conscience. The way in which we concluded the chapter in which we describe the concept "culture" seems to be appropriate for the beginning of the presentation in this chapter. Constantin Noica asserted a grim truth saying that we are unsatisfied. We quoted those lines in order to underline the fact that this dissatisfaction may also be translated by our need of (re)defining ourselves more clearly based on our culture and our conscience as people. The dissatisfaction also appears as a motif at Lucian Blaga which makes the analogy between culture and this condition of the man trying to determine what the culture means. Thus, "one could not say that at the basis of the culture would be an alleged dissatisfaction of the man towards the immediate moment and a need to escape in any way from its web and nothingness (...). The dissatisfaction towards the immediate moment is rather a secondary fact, a derived aspect which occurs exactly because the man feels he is meant to have another destiny than the one of the existence in and for the immediate moment" (Blaga, 2011, p. 459). The dissatisfaction is not only related to what we could have been and we are not as individuals or nation, but it is exactly the basis of the cultural production and of the understanding of our worldly existence.

The Romanian philosopher adds the fact that "existing as a man means from the start finding a distance from the immediate moment by placing in the mystery. The immediate moment only exists for the man to be superseded. The immediate moment only exists for the man as a passage. (...). By his revealing trials becomes however a creator, namely a creator of culture generically. (...) From this point of view, the culture is not a luxury that the man affords as an adornment, which can exist or not; the culture results as a complementary issuance from the specificity of the human existence as such (...). The existence enriches with its most profound version. The culture is the visible sign, the expression, the figure, the body of this existential version" (Blaga, 2011, pp. 459-460). This superseding of the immediate moment reminded by Blaga is exactly the cultural creation which determined the appearance of what we call nowadays cultural landscape. We must retain the reference to the fragment, since we will use this aspect of the immediate moment and of the everyday life in the chapter dedicated to the chronotopy of the cultural landscape. However, for the moment, it is enough to understand the culture as a visible expression of existence as pointed by Lucian Blaga. It is the essence of the term "cultural landscape" as we will deal with in this paper, from the point of view of the geographic paradigm, namely the understanding of the concept in the form in which it appeared.

If we were to make a short presentation of the way in which the cultural landscape is understood, we will remember the most important aspects that we motivated in this first part of our thesis. Therefore, the cultural landscape is seen as an uninterrupted process of construction (Ingold, 1993) which is (re)created based on the existence of a temporality. This temporality, as we will deal with it within our following chapters, is at the origin of its existence in the form of a chronotope. But we will discuss about this aspect a little bit later. The cultural landscape is an object of conscience, a heterotopia, a mirror of the society. It is at the same time an expression of cultural values and a symbol (Backhaus, 2009), as well as the complex expression of a life style (Sârbu, 2011), as result of the interaction of cultural factors with the natural ones which determine together a specific structure with a strong symbolic significance. The cultural landscape is also a moral order (Ezra Park, 1926) or a landscape of power (Zukin, 1991). It is a cultural process (Hirsch, 1995) which supposes the "uncertain coexistence of a plurality of the partial order" (Palermo, 2008, p. 32) being a space of relations and of inflexion (West-Pavlov, 2009) which requires to know how to read between lines, such as a text. The implication is determined based on the way in which we know how to read and understand the text, namely the cultural landscape, because it "juxtaposes or overlaps the <<the traces>> of several previous periods whose primary meanings we miss since every trace is reinterpreted and reintegrated in the contemporary functional context" (Mihali, Copoeru, 2002, p. 11).

7.5.1 Conclusions on the understanding of the concept as metaphor and palimpsest

Considering these brief descriptions of what we enlarged in this entire chapter, we think it is appropriate to emphasize the existence of the cultural landscape in the form of a metaphor (Jackson, 2003) and a palimpsest (Calcatinge, 2011; Schein, 2010 et al.). For this purpose, Lucian Blaga asserted the fact that "unlike the substance of real things in the sensible world, the substance of creations does not possess significance and sense by itself: here the substance always substitutes something else; here the substance is a sediment implying a transfer and a conjugation of terms belonging to some different regions and areas. The substance hereby acquires a so-called <<metaphoric>> aspect" (Blaga, 2011, pp. 348-349). The Merriam-Webster Dictionary defines the metaphor as "a figure of speech in which a word or phrase literally denoting one kind of object or idea is used in place of another to suggest a likeness or analogy between them"[75]. This is why we consider that this way of relating with the cultural landscape in the form of a metaphor make its meaning to be extremely varied. It is equally a cause and an effect. Essentially, regardless of the definition or meaning that we give to the term, it will refer more or less to the same elements and categories, however with small variations. Even if there are thinking schools referring to the cultural landscape as being what was transformed by the man or if they refer to the agricultural landscape or only to landscapes with a certain symbolic significance, thus regardless of these varied references, we consider that the term and concept "cultural landscape" will be a metaphor by means of the fact that it will always give a new significance keeping at the same time a single meaning to which it will be constantly compared.

We need to add to the elements we mentioned above an affirmation of Lucian Blaga who thought that "in the genesis of metaphor we have to see a burst of the human specific in all its spread. The metaphor (...) is limited, a spiritual function, to those resulting from the conditions, beyond the vicissitudes of time, of its genesis. 1. It is convoked either to compensate the insufficiencies of the direct expression for an object or 2. to reveal hidden, real or imaginary parts and significances of an object" (Blaga, 2011, p. 358). In the same order of ideas, Annamaria Lossi argued "the creation of new metaphors opens new possibilities for the formation of meaning, it discloses new rules in linguistic behaviour, rules related to the symbolic sphere. Accordingly, the concept of symbol has to be related to that of metaphor" (Lossi, 2010, p. 211). Thus, it is admitted the creative dimension of this figure of speech, because "metaphors describe a reality outside of reality itself. As etymology indicates, a metaphor transfers us from an <<old>> meaning to a <<new>> one allowing a different understanding. The function of metaphors is to convey attention to something different in order to be able to understand how something was given" (Lossi, 2010, p. 213).

As regards the existence of the cultural landscape in the form of a palimpsest, a series of authors have given their opinions on this aspect. In the final chapters we will get back to this term and in this chapter we will only make a short presentation of this way of understanding the cultural landscape. The Merriam-Webster Dictionary defines the term "palimpsest" as "writing material (as a parchment or tablet) used one or more times after earlier writing has been erased" or "something having usually diverse layers or aspects apparent beneath the surface"[76]. Thus, the cultural landscape as palimpsest is the proof of consecutive interventions due to the man on the environment. Each of these interventions is noticed either more strongly or more weakly, but the traces of each of them are visible. In this regard, Alexandru Dragomir asserted the fact that "palimpsests show that in life (in the past, in the History) we do not encounter an enlistment of ends and beginnings but with an interlacing of such

[75] Definition given by the Merriam Webster Dictionary. Website http://www.merriam-webster.com/dictionary/metaphor accessed on 28.09.2011.

[76] Definition given by the Merriam Webster Dictionary. Website http://www.merriam-webster.com/dictionary/palimpsest accessed on 28.09.2011.

series (<<beginnings-development-end>>), one on top of others. (...) In case of the palimpsest, there are two realities, one on top of the other: one covers the other one but not perfectly, not irretrievably, not even enough that the underneath one be over" (Dragomir, 2008, p. 245). This palimpsest, besides the fact that it cannot be seen as a succession of human interventions, may be seen as an overlap of the natural with the anthropic (or cultural) element. The natural element is the one underneath which as Alexandru Dragomir mentioned above, is imperfectly covered by the anthropic one. This does not denote a finality of the first one. It is still visible, it is the first layer of the palimpsest, on top of which layers of the human interventions were added and suppressed. Are these layers of "what does the man create? Myths, legends, ambiance, philosophy, science, arts of all types (poetry, prose, picture, music, sculpture, architecture), technique, politic, logic. Some answer to needs, other develop these needs and others (?) are free of charge. We are now in a humanized world (in an <<environment>>) in which the technique and not the nature is the stranger (the object); a world in which <<nature>> is relegated (reservation, place of recreation, zoological gardens)" (Dragomir, 2008, p. 198).

The relation palimpsest-metaphor is realized by means of a narration of the landscape that we will mention in the chapter dedicated to the chronotopy and to the one dedicated to its existence as Weltanschauung. The construction of this relation is also prefigured by Constantin Noica who asserted the fact that it is about a mixture of "known and unknown which makes the literature (...) which makes music, landscape, knowledge" (Noica, 2012, p. 23). The unknown mentioned by Noica is the creating act, the human intervention based on thinking, experience and conscience. But here too, we can extrapolate asserting the fact that creating act itself is a mixture of known and unknown. These two components are at the basis of the narrative construction of the cultural landscape and of the relation palimpsest-metaphor. We will detail however this aspect in the following chapter dedicated to the contributions brought to the theory of the concept "cultural landscape". In this chapter, we only want to briefly conclude the way in which we relate with the concept at the level of its understanding.

7.5.2 Conclusions on an operational definition of the concept

Give what we have presented so far, we will give a working definition that we built based on the chronotopy of the cultural landscape and that will reappear in the respective chapter in a slightly changed form. We do this in order to underline the way in which we understand the cultural landscape and to prefigure the importance of chronotopy and of its existence as Weltanschauung analyzed in our thesis. Thus, we can assert the fact that the cultural landscape, when shaping its existence in the form of an ideal chronotope (palimpsest) and of a social dialogism (metaphor) is the spatial representation of the succession (here comes the temporality) of interactions and social relations determined based on cultural processes. However, we have to stop for a moment on this aspect and to understand the term "cultural process". According to the Longman Dictionary of Contemporary English, the term "process" is defined as "a series of actions that are done in order to achieve a particular result" or "a series of things that happen naturally and result in gradual change"[77]. At the same time, we have reminded earlier the fact that the cultural landscape itself is understood as a cultural process (Hirsch, 1995). Can this term be defined by means of other cultural processes? How do we describe these cultural processes? In the line of the elements mentioned so far in this entire chapter, we will describe cultural processes by means of what is known to be cultural creation, because "seen as conscience fact, the culture represents its main, decisive side: all these crystallised contents of conscience, more or less elaborated, in the form of knowledge, representations and notions, scientific concepts and theories, literary, artistic images, moral values, the creating process that led to their appearance" (Tănase, 1968, p. 285). Lucian Blaga said related to the cultural creation that "it is thus an invention of the human spirit, an invention of metaphoric nature." (Blaga, 2011, p. 410). This cultural creation, from the perspective of processes by means of which it appears, as we asserted in the given definition, is at the origin of some interactions and social relations. They are governed by the human experience and by the conscience, as we argued in the previous chapter. The cultural creation may also be based on a culture which can be understood as asserted Constantin Noica "as an endless medicine. It is not the culture as <<care>> of the man for himself, but the culture as detection of diseases of the real world of any type, with no attempt of its recovery. You only bring forward the insecurities of the real world – and you give them names" (Noica, 2008, p. 224). Wilhelm Dilthey said in this line the fact that an understanding of nature and of our social existence can only be performed by "searching and reflecting on the unobservable content which expresses itself in the exterior, observable course of human events" (Dilthey, 2001, p. 34). As one will notice in the last chapter of the paper, this is where originates the understanding of the concept and theory governing it.

Therefore, the definition proposed for the concept brings together in a metaphoric sense all elements that we considered indispensable in the theoretic study of the cultural landscape. We can thus conclude that, although we cannot support the study of the cultural landscape by a single holistic definition, it is necessary to bear in mind the formulation of an operational definition specific to the research of each of them, that should answer at the same time to the purpose of the study and to a clear and easy understanding. This definition must answer at least to principles of independence and intellectual curiosity remind-

---

[77] Definition from the website http://www.ldoceonline.com/dictionary/process_1 accessed on 29.09.2011.

ed in the subchapter dedicated to the research method. Regardless of the way in which we relate at the theoretical level with the cultural landscape, we consider that a minimum of exigency must exist in the explanation of the concept which is to be studied by establishing, if not a clear definition, at least some structuring principles that would help the reader to understand on what we will focus in the study out of everything that could mean the cultural landscape. It is superficial and maybe too vague to begin a study in which we would constantly discuss about the cultural landscape without clearly delimitating from the beginning what we will focus on both at the theoretical and at practical level. And for this purpose, this thesis wants to propose a way of understanding the need of assimilating the theory to such a complex and at the same time ambiguous concept. This happens because the study of culture and landscape requires the existence of a culture of the study itself; we cannot speak about the cultural landscape without knowing the concept, its meanings, the definitions given over time and more importantly without relating it to our everyday existence, to the reality in which we are meant to live. In other words, we consider that the study of the cultural landscape, like any other study within human sciences – such as architecture and urbanism, is based on the triad understanding-experience-knowledge (not necessarily in this order). Thus, Tudor Vianu said that "the so-called <<experience>> proceeds by the assimilation of the present and future with the past and consequently of the unknown with the known" (Vianu, 1982, p. 464). However, "although the experience puts us in front of the reality of life in its multiple ramifications, we seem to know a single thing namely our own life. It is the knowledge of something unique and no logic help can beat this limitation (...). Only the understanding supersedes the limitation of the individual experience and at the same time gives to the personal experience the character of knowledge of life" (Dilthey, 2001, p. 38). Thus, the understanding "supposes experience, and the experience becomes knowledge of life only if understanding leads us from experience limitation and subjectivity to entirety and generality" (Dilthey, 2001, p. 38).

8 Role of Theory in the Study of the Concept of Cultural Landscape

8.1 Theoretical perspective on the pluridisciplinary study of the concept

Ever since its first appearance in the academic discourse in 1908, the term "cultural landscape" has fed with no interruption the need for knowledge and at the same time has built and destroyed thinking schools, ideologies, paradigms or theories during the entire 20th century and even more assuredly in the 21st century. A study on this aspect is the one carried out by Urszula Myga-Piatek (2008) who suggests an analysis of the way in which cultural landscape studies developed during the existence of the concept. The author reminds a "human paradigm" (Myga-Piatek, 2008, p. 84) in cultural landscape studies which is based on the existence of an alleged unity between nature and culture. Therefore, "Besides methods which are typical of natural sciences, the paradigm, being the determiner of the anti-scientistic orientation, also adopts methods of humanistic and social sciences in cultural landscape analysis. This approach makes it possible to study immaterial and mathematically immeasurable elements of the anthroposphere and define cause-result relations occurring between the nature and the culture" (Myga-Piatek, 2008, p. 84).

At the same time, we can notice even from what we have argued so far in our thesis that the research process of the cultural landscape has known a series of key points during its evolution. The concept itself has always had the necessary vitality in order to maintain its fascination with which it constantly attracted the interest of researchers and "currently, great opportunities appear for cultural landscape research in terms of studies and application. (...) The main aim of modern research is detection of the mechanism of changes of elements of landscape, and, basing on this recognition, assessment of the pattern and tendency of transformation" (Myga-Piatek, 2008, p. 86).

These opportunities in the research of the cultural landscape occur in a pluridisciplinary framework and an important thing we must not omit is that within some scientific and academic studies, the creative part and the contribution to knowledge must be exercised unequivocally, thus contributing to possible developments of some already existing theories or to the appearance of new ones. Thus, we think that the concept of cultural landscape would not have known the current evolution if it were not launched, studied and developed in the academic and scientific environments, thus creating a way in order to be used even in political discourses. However, in this chapter we will not focus on the study of political discourses but on the need to train specialists by pluridisciplinary university programmes and by publishing scientific papers in the field. This pluridisciplinarity is essential in the contemporary studies of cultural landscape and "while discussing the scope of modern applied research of cultural landscape, one could observe its <<dangerous close approach>> to other disciplines and directions, including the architectonic aspect. As the result, cultural landscape is becoming a <<competitive>> notion (...). This inspires searching for common initiatives" (Myga-Piatek, 2008, p. 87).

Analysing the subjects involved in the cultural landscape study, in order to correctly fit the object of our research as architects and urban planners as well as to find the place that we hold as professionals in such studies, we considered it was appropriate to present the area in which the architecture and urbanism fit into. Thus, knowing the fact that we operate an important concept in geography, we could notice in our research that many theoretical aspects are tributary to this associated subject. A presentation in this line was also made by Myga-Piatek (Myga-Piatek, 2008, p. 76) who showed within a scheme the way in which is divided the interest for the cultural landscape study according to the involved subjects. Thus, the architecture, the urbanism and the landscaping have been fitted into a practical field and in the theoretical field have been included the cultural geography, the human and the regional one, ethnology or philosophy. The practical field, together with the above mentioned subjects, sociology and psychology have also been included.

In this respect, architects, urban planners and landscapists are among the specialists which should have a thorough (and correct) training as regards the theoretical study of the cultural landscape, before dealing with the practical part of studies. This is necessary in order to be able to have a professionally valid discussion with colleagues involved in the process of cultural landscape study. We cannot eliminate the importance that the theory has even within some creative subjects such as architecture, urbanism and landscaping. We know the fact that the entire creation of architects (landscapists) and of urban planners is (most of the times) based on a strong theoretical and philosophical foundation. In this order of ideas, their contribution must be distinctive, we think, both in terms of theory and practice within these studies. As regards the practical side, the situation is slightly different, since we are the creators (in the past, present, future) of what others only have the (humble?) privilege to study from an external position. At the theoretical level however, we have imported and used this concept many times too brutally in our profession. But leaving aside its name and the fascination it presents, we consider that we have at hand, as quickly as possible, the baggage necessary for the operation at the theoretical level of this concept in the autochthonous academic and professional environment, based both on contributions of colleagues geographers, anthropologists, historians or ethnographers, and on the contributions specific to our profession. The latter ones have to weigh more in order to delimitate the field in which we will operate this concept. Thus, Frederick Aalen (2004) suggested a temporal analysis of subjects dealing with the cultural landscape study (Aalen, 2004, p. 14). On the temporal axis also appear the architecture, the urbanism and the landscaping, but in a presentation on the implication of landscape studies within different subjects, the author does not classify anymore the three above mentioned subjects (Aalen, 2004, p. 14).

Our work being a predominantly theoretical one has focused with no interruption on this side of the theoretical study of the concept "cultural landscape" that we consider it should be developed from the perspective of the interest it presents for the architect and the urban planner. But having at hand too few theoretical approaches of this type proposed by specialist in the autochthonous academic environment, we will refer to reference works already known in the foreign academic environments.

In the construction of the arguments within our thesis, we referred to many of these theoretical papers but even so, what we can propose in this chapter will essentially be the succinct presentation of the way in which the cultural landscape is studied and of the methods on which the concept development is based. This aspect will contribute to the consolidation of our theoretical argument built in the entire thesis. We will however pay attention to the autochthonous specialty literature in order to emphasize, even in a criticism-based way, certain approaches that have been realized in cultural landscape studies.

From the numerous volumes read for the construction of our arguments on some theories of cultural landscape study, we have chosen for an extended analysis in this field of research methods only a part of them, the ones that we considered to be able to bring an increased contribution to the way in which architects and urban planners should relate to this topic. At the same time, we must take into consideration that in this way we do not minimize the contribution of the other authors to the discussed topic but we only show the vast theory that accumulated around this concept which most of the times is avoided by specialists or is simply ignored.

8.1.1 Philosophical approaches in the study of the concept

In its attempt to determine the importance the theory has in the theory of geography, Phil Hubbard suggests a solid argument of the way in which we should relate to different aspects of life especially as regards the study of some complex operational concepts such as the "landscape" or the "cultural landscape". Generally, we can admit the fact that this entire knowledge process that we carry out predominantly in academic and scientific environments, has the role to explain events that define our everyday life. Therefore, in an inevitable way it is reached the knowledge, the elaboration and the use of life philosophies. These philosophies have however the role to support the theoretical discourse which should virtually be at the basis of a good practical initiative. In the cultural landscape study, both in geography and in architecture and urbanism, it has almost constantly interfered the matter of determining valid theories on the space which led to the appearance of a "more meta-theoretical project, a search for an appropriate ontological and epistemological location for spatiality, an active <<place>> for space in a Western philosophical tradition that had rigidly separated time from space and intrinsically prioritized temporality to the point of expunging the ontological and epistemological significance of spatiality" (Soja, 1989, p. 119). Consequently, "in the simple sense, philosophy can be described as being theory about theory. It advances a fundamental and general analysis on life (...). Just like there are many theories in geography, at the same time many definitions exist on and ways of approaching the philosophy. Philosophical interrogations refer to the fundamental aspects of life: reality, knowledge nature, mind, matter, truth and so on" (Hubbard et. al., 2005, p. 4). Thus, four components are provided to us as regards the way in which philosophy must be analyzed in the field of cultural landscape study in geography. They are ontology, epistemology, ideology en methodology (Hubbard et. al., 2005, p. 5). In this chapter, we will only refer to the first two of them, to which we will add the phenomenology.

8.1.2 Ontology in cultural landscape studies

According to Longman Dictionary of Contemporary English, ontology is a subject of study in philosophy that is concerned with the nature of existence. Thus, Ian Graham Ronald Shaw, Deborah Dixon and John Paul Jones III connect the term "geography" with the term "ontology", in order to determine the way in which the two of them can work one for another. It is known that "it concerns the set of specific assumptions about the nature of existence underlying a theory or system of ideas; beliefs about what exists and can be observed, and therefore known" (Hubbard et. al., 2005, p. 5). These assumptions and beliefs about the existence features, as authors affirm, are determined for the knowledge of world in which we live and which consists of different components and subcomponents which can be classified and included into different categories. Thus, two categories which can be distinguished are the natural and the social ones, and regardless of the way of their division in various components, all are characterized by a specific duality. The authors thus classify in a table (Shaw, Dixon and Jones, 2010, p. 11) the ontological components of the world, which result among others in duality between reality and fantasy, body and mind, nature and culture, oneself and the other, organic and technologic or space and time. These above mentioned components are the only ones that we will analyze in our thesis.

This classification facilitates, for instance, the understanding of the way in which certain connections determine specific relations between the above mentioned dualisms. The authors explain thus how occurred the geographer's perception change as regards the way in which they relate to the studied elements (Shaw, Dixon and Jones, 2010, p. 12). Ontology has determined the separation of geographers according to their attitude towards the way in which they relate to the study of existence. Consequently, some of them have separated the analysis from the dominantly material and historical part in order to study the symbolic and immaterial part, representing those who "postulate the fact that values and opinions may be known either by scientific research methods or by other means, thus being legitimate constituents of knowledge production" (Hubbard et. al., 2005, p. 5). However, at least until the middle of the last century, many remained tributary to the scientific, quantitative, material analysis. On this occasion, the spatio-temporality has also changed its coordinates, according to which it was studied, acquiring an increased importance in the analysis of geographers keeping at the same time the strong philosophical significance. The understanding of time and space determines the understanding of the causality of the elements composing the environment (Shaw, Dixon and Jones, 2010, p. 12) and "this view of space enables us to describe things as being near or far away from each other; it also leads to a number of fundamental geographic concepts, such as absolute and relative location, distance and direction, scale and movement, as well as perspective" (Shaw, Dixon and Jones, 2010, p.12). In this respect, Edward Soja stated that "tied to this interpretation of the connections between physical, mental, and social space is a key assumption about the dynamics of spatiality and hence about the relations between (social) space and time, geography and history" (Soja, 1989, p. 122). Thus, ontology helps understanding the processed taking place spatially, the transformations which led to the definition of our existence over time and the way in which we relate to the space. In conclusion, Phil Hubbard, asserts the fact that "different ontologies informing geographical theory therefore set out a number of positions such as: the things we experience are the things that exist; the things we agree amongst ourselves to exist, exist; what exists is what is perceived to exist; and so on" (Hubbard et. al., 2005, p. 5).

8.1.3 Epistemology in cultural landscape studies

Epistemology deals with the way in which we understand knowledge (Shaw, Dixon and Jones, 2010, p. 15) from the perspective of science or "means the way in which knowledge is obtained or imposed; assumptions on the way in which we can know the world", (Hubbard et. al., 2005, p. 5). Thus, in a dualist world such as the one in which we live "the central but somewhat vague definition of epistemology – assumptions and theories about <<how we know the world>> - can be more usefully stated as an understanding of how we exist as knowing subjects, how we actively engage with the world, and how we reflect upon our experience of the world" (Shaw, Dixon and Jones, 2010, p. 15). According to Merriam-Webster Dictionary, the epistemology is the study or a theory of the nature and grounds of knowledge especially with reference to its limits and validity, and in the study of geography and of other sciences it makes available a wide range of knowledge types. In this regard, the authors have determined a series of paradigms as regards the research types in geography. The synthesis they realize (Shaw, Dixon and Jones, 2010, pp. 20-22) helps understanding the way in which each of these paradigms may be assimilated with knowledge.

Therefore, four types of paradigms are presented to us: (1) the paradigm of spatial sciences (which emphasizes the spatial relations); (2) the paradigm of humanism (in which places are interpreted by their meaning); (3) the paradigm of critical realism (in which structures and mechanism behind the events are analyzed); (4) the paradigm of post-structuralism (in which the production and the import of the spatial discourse are analyzed). As the authors suggest, all these paradigms, some of them being analyzed more detailed in our thesis, propose a mode of relation to the way in which "the research process is always <<in the middle>> of our own experiences and efforts" (Shaw, Dixon and Jones, 2010, p. 19).

Moreover, for each of the above presented paradigms, the authors determine specific research methods based on the theories which gravitate around each of them. Thus, for the paradigm of spatial sciences, the defining theoretical framework is specified regarding the way in which the space is considered. It is a bi-dimensional one in which places have distinctive, measurable and quantifiable attributes, emphasized by an exact mapping. The way of approaching the study supposes an objective attitude of the researcher towards the studied object and the way of storing data is based on quantitative, rigid analyses (socio-economic, demographic, types of vegetation etc.) (Shaw, Dixon and Jones, 2010, p.20, table 2.4). In case of the second paradigm, the humanist one, it is clearly distinguished the way in which the mental and symbolic part dominates the material one. The qualitative governs the quantitative, reference being made to terms such as genius loci or sense of place (detailed at the same time in our study). In this case, the subjective part dominates over the objective one, personal interpretations of researchers being also accepted, based on the experiences lived by the latter, but this aspect is specific to the phenomenology. Data analysis supposes the studying archives, traditions, the history and the experience of individuals in the analyses area, data of ethnographic nature, cultural specific, language (Shaw, Dixon and Jones, 2010, p.20, table 2.4). Briefly, it is that thick description also found in the study of the architectural landscape at Iain S. Black. It is applicable at the same to the landscape rurality presented concisely by Keith Halfacree. The third paradigm, the one of critical realism supposes a detailed analysis of social structures and of different mechanisms which are found at their basis, which supposes the need for the analysis of obtained data by direct observation, interviews, direct sources. This is based on a complex situational knowledge articulated to the objective forces and structure of social mechanisms (Shaw, Dixon and Jones, 2010, p. 21, table 2.4). The last paradigm, the post-structuralist one, supposes the visual and textual analysis of data, the analysis of binary structures presented in the part of ontological, seeking to determine total knowledge both on the studied object and on the researcher himself (Shaw, Dixon and Jones, 2010, p. 21, table 2.4). Relations between individuals and instruments are materially determined by discourses which take place at social, political or economic level, which supposes the existence of a complex spatial construction based on power. Hence, by means of what we have stated so far, one can easily distinguish differences that exist at the epistemological level between human and physical geography. In case of the cultural landscape study, which is a branch of the human geography, the theories specific to it apply, "hence, epistemology in human geography is often predicated on interpretative frameworks adapted from the humanities rather than the natural sciences" (Hubbard et. al., 2005, p. 6).

8.1.4 Phenomenology in the study of cultural landscape

According to Gabrielle Hezekiah, "the phenomenological attitude recognizes the subjectivity of all experience and seeks to make that experience the focus of an investigation. In Husserl's investigation, this involves an appreciation of the fundamental role of consciousness in our perception of the world" (Hezekiah, 2010, p. 7). Thus, at the phenomenological level, the main aspect is determined by the fact that "the information related to the surrounding world is obtained by the observation process which is essential in the knowledge production" (Rhoads and Wilson, 2010, p. 27). A supporter of this humanist phase in the geographic studies was Yi-Fu Tuan, who emphasizes in its studies the way in which we understand the environment, in which we perceive it and in which we report to it, in which we succeed to produce plus-value. Thus, in human geography, space perception and "observations, then, are dependent on the observer's relative position and, rather than being straight-forward perspectives of a world that timelessly communicates its essence through easily created categories, are a complex product of human experiences" (Rhoads and Wilson, 2010, p. 35). One of the important aspects of phenomenology, with direct implications in study methods of geography is the one of significance. In our research we have granted quite a great importance with ramifications in architecture and equally in urban studies, since "phenomenology is a philosophy which takes into consideration the individual experience – but with the ultimate purpose of producing a solid basis for knowledge, greatly due to emphasis on perception and knowledge" (Hale, 2011, p. 111). We should thus remember what we stated in the chapter referring to the appearance and the evolution of the concept "cultural landscape". We underlined the fact that phenomenology has appeared in the study of geography in the years ’60 and ’80 of the 20th century (Shields, 1988, p. 29). Rob Shields named particularly this phenomenology as one specifically geographic since “geographers adopted a <<phenomenological stance>> to oppose the extension of rational scientific analysis” (Shields, 1988, p. 30). Considering also what we stated in the first chapters of this thesis regarding the understanding of the cultural landscape in the form of a text, we can continue the argument by reminding what Edward Casey asserted: “what distinguishes the literary work from a scientific theory is its peculiar structure: every work of literature presents itself to the reader as a series of strata that, though organically bound together in the reader's experience, are distinguishable upon analysis” (Casey, 2010, p. 2).

Therefore, even at this level, of the landscape assimilation with a work of literature which must be read, understanding the strata (we should remember the term “palimpsest”) of the cultural landscape supposes the appeal to the human individual experience. But this thing is not the only one important since within the phenomenological study of the landscape, besides the human experience, other aspects of perception and conscience are involved, “thus phenomenology involves the description of things as one experiences them, or of one's experiences of things (...). One important class of such experiences of things is perception – seeing, hearing, touching, and so on. But it is by no means the only one. There are also phenomena such as believing, remembering, wishing, deciding and imagining things; feeling apprehensive, excited, or angry at things; judging and evaluating things; the experiences involved in one's bodily actions (...)” (Hammond et.al., 1991, p. 2). Nevertheless, a purely phenomenological approach within the cultural landscape study is less probable, since “presuppose a unique datum of human experience which remains unestablished in his arguments, against the by now quite important weight of psychological, linguistic, and ethnographic evidence which supports the postulate of a multiplicity of different frames of reference tempered only by the force of normative socialization which structures and <<frames>> experience for us” (Shields, 1988, p. 39). Hence, we have to admit the strong phenomenological structure of the process of cultural landscape study and this aspect is exactly the basis of our argument in this thesis. We can also provide the conclusion by what Rob Shields stated regarding the phenomenological study in geography. Thus, “geographical phenomenology is a combination of idealism (in the primacy given to the individual agent), existentialism (especially in the notion of intersubjectivity discussed below) and phenomenology (in the notion of lifeworld)” (Shields, 1988, p. 31). Here a series of additional explanations have to be offered. We can notice the fact that in the original quotation appears the term lifeworld, term that we associated to the term “Fact-of-being-in-the-world” or in German *In-der-Welt-Sein* of Martin Heidegger, a term which is connected to *Lebenswelt* (lifeworld) used by Edmund Husserl and to the term *Lebenszusammenhang* used by Rob Shields in the description of that “geographical phenomenology” (Sheilds, 1988).

8.1.5 Methodological study of the concept of cultural landscape

First of all, in this chapter we will refer to discursive studies, to subjectivity and situational knowledge as research types within cultural landscape studies. Thus, we must remind the fact that discursive studies are related to post-structuralism, ideologies being directly designated in the language by the narrative structure of discourse which influences the way in which the individual perceives the surrounding objects which are related to different theories of power. Thus, we find the type of discourse described in the work of Foucault (1988). This discourse must be seen as a singular, particular one, as well as a general one. In the acceptance of the general meaning, the discourse consists not only of words but of ensemble[78] (even architectural ones), understood as heterogeneous including institutions, buildings (architectural forms), that we today call urban landscape, legislative system, philosophical discourses and arguments (Foucault, 1988, pp. 194-195)[79]. Architectural ensembles may be spatial units or groups of buildings, dwelling units, spaces and places with cultural connotations which create in the end a cultural landscape. As such, the discursive landscape is the one that produces a certain communication way, a language in which power relationships are converted into urban texts.

What is real and how we know if the opinion we form ourselves on the real is true is a question that has preoccupied the researchers for a long time. In order

[78] Original quote:"What I'm trying to pick out with this term is, firstly, a thoroughly heterogeneous ensemble consisting of discourses, institutions, architectural forms, regulatory decisions, laws, administrative measures, scientific statements, philosophical, moral and philanthropic propositions–in short, the said as much as the unsaid. Such are the elements of the apparatus. The apparatus itself is the system of relations that can be established between these elements. Secondly, what I am trying to identify in this apparatus is precisely the nature of the connection that can exist between these heterogeneous elements. Thus, a particular discourse can figure at one time as the programme of an institution, and at another it can function as a means of justifying or masking a practice which itself remains silent, or as a secondary re-interpretation of this practice, opening out for it a new field of rationality. In short, between these elements, whether discursive or non-discursive, there is a sort of interplay of shifts of position and modifications of function which can also vary very widely. Thirdly, I understand by the term "apparatus" a sort of–shall we say–formation which has as its major function at a given historical moment that of responding to an urgent need. The apparatus thus has a dominant strategic function. This may have been, for example, the assimilation of a floating population found to be burdensome for an essentially mercantilist economy: there was a strategic imperative acting here as the matrix for an apparatus which gradually undertook the control or subjection of madness, sexual illness and neurosis" (Foucault, 1988, pp. 194-195).

[79] The initial Foucault's essay entitled The confession of the flesh was written but was not published before his death. What was published in the volume edited by Colin Gordon (1988) entitled Power/Knowledge: selected interviews and other writings, 1972-1977 by Michel Foucault is an interview from 1977 which has this title. In the interview, Foucault explains the way in which he understands the terms "apparatus" and "ensemble".

to be able to verify the scientific argument, there is need generally for a series of empirical data based on the observation and investigation of the external environment. The simple relation of the type object-text is replaced by another one more complex in which both the researcher and the audience, the spectator, is given a role in the process of decoding the language, the text of the landscape, since "our ability to understand relies on our capacity to interpret signs. It also presupposes that signs come to signify in ways that no particular author or speaker can constrain in advance through intention" (Murdoch, 2006, p. 8).

The situational method experienced a spread following the studies carried out by researchers in the years '80s of the 20th century, in the cultural momentum produced by sociology, anthropology, geography and cultural studies: "The main point they make is that academic and other ways of knowledge are always situated, always produced by positioned actors working in/between all kinds of locations, working up/on/through all kinds of research relations(hips)" (Cook, 2005, p. 16). Iain Cook arguments clearly the connection between the "text" and the "object of study" for the purpose of explaining the importance of situational knowledge and of the used terms. Thus, he states the fact according to which the academic knowledge is situated and at the same time the involved actor, the researcher, is always positioned: "thus, so the argument goes, writing about academic knowledge as a relational process rather than a straightforward thing might highlight the politics of knowledge in academic research, produce more modest, embodied, partial, locatable and convincing arguments and, in the process, make it possible for researchers (and their audience) to see and make all kinds of, often unexpected, politically progressive connections" (Cook, 2005, p. 16). Then he explains that it is this state of facts, this existence known and clearly positioned in an active reality is the one that leads to understanding processes and to decoding objects of study which can equally be a text, a language such as the one designated by the landscape, in the built environment and their transformations into texts.

Briefly, we can state the fact that all around us there is knowledge. It exists in different modes, encoded and most of the times hidden and waiting to be decoded. Thus, the situational knowledge is specific to a certain situation through which the spectator has a certain piece of information and that he/she uses for a certain purpose. Subjectivity, which is tightly related to the experience of everyone and to the way of learning, leads to an advanced situational knowledge which is based on culture and traditions but also on the language.

It is necessary here to remind (again), in the construction of our argument that a knowledge based on the experience is called a posteriori but there is at the same time a priori knowledge, which Immanuel Kant considered transcendental knowledge while the a posterior was called empirical: "The possibility of synthetic propositions a posteriori, i.e., of such as are drawn from experience, also requires no special explanation; for experience itself is nothing other than a continual conjoining (synthesis) of perceptions. There remain for us therefore only synthetic propositions a priori, whose possibility must be sought or investigated, since it must rest on principles other than the principle of contradiction" (Kant, 2004, p. 27). A priori knowledge types are extremely rare but among them we can mention time and space knowledge. A posteriori knowledge refers to researcher's identity and practices which represent his experience but which does not guarantee the complete decoding of the hidden code of the object of study, thus the understanding being a partial one and the knowledge being incomplete, the individual will have to know the studied object by a limited number of pieces of information.

Thus, in order to initiate any kind of valid study for the purpose of knowing an object or phenomenon, there is a need for a method or several combined methods which apply based on a theoretical knowledge better substantiated in advance. According to Phil Hubbard, the "methodology is the set of procedures used to develop or test a theory" (Hubbard et. al., 2005, p. 6). At the same time, it is "a body of methods, rules, and postulates employed by a discipline: a particular procedure or set of procedures" or "the analysis of the principles or procedures of inquiry in a particular field" according to Merriam-Webster Dictionary. Thus, all research methods presented in this chapter, either specific to geography or imported from other subjects, compete for the emphasis of methodologies specific to the cultural landscape study. It is important to remember under the circumstances the fact that "these procedures must not deviate from the ontological and epistemological assumptions of the overall philosophy of investigation" (Hubbard et. al., 2005, p. 6). At the same time, Phil Hubbard emphasizes the changes that occurred among the research traditions in geography where philosophy has earned and increased importance, which led to a diversification of theoretical methods and approaches. Even so, "Many geographers doubt that philosophical issues are actually relevant to geographical research. [However] no research (geographical or otherwise) takes place in a philosophical vacuum. Even if it is not explicitly articulated all research is guided by a set of philosophical beliefs" (Michael Hill quoted in Hubbard et. al., 2005, p. 7). Here is what Pamela Shurmer-Smith said about the need of knowing the theory: "the way to understand theory is to start by recognizing that everyone is (always and already) using it to construct meaning. (...) Throughout the world, there are people who specialize in bringing latent theory to the surface; obviously academics come into this category, but so do religious specialists, artists and politicians. All of these to some extent reify theory – make it into a thing in itself, lever it apart from its ordinary implication within life" (Shurmer-Smith, 2002, p. 11). All the more so as the study began by us refers to the cultural landscape, hence to the cultural component, we can add the fact that "there is no outside position when theorizing culture, not even when one is trying to understand the ideas and values of <<others>>. Culture is inherently theoretical, and to comment upon it is to theorize about theory. It requires that one becomes self-reflexive, conscious of one's own viewpoint when trying to understand viewpoints in general" (Shurmer-Smith, 2002, p.

12).

We consider that the milestone for any incursion in a subject of study such as the cultural landscape is the well knowledge of theories governing it. This is all the more important as we operate an imported and adapted concept, coherent or not, to the needs of our profession. Within our research, the need to know as many things possible on the object of study has led us, as expected, towards a pluridisciplinary study. Thus, the book edited by Basil Gomez and John Paul Jones III entitled Research Methods in Geography. A critical introduction has provided us precious information on the way in which the concept "cultural landscape" is analyzed in the subject which initially launched it.

The interest of the above mentioned authors is to facilitate both for students and for any person interested in the study of geography, a set of recognized methods and techniques, as well as analysis of theories which have governed for centuries some branches of the subject. The volume begins with a valid observation which is alarming at the same time and which warns on the way in which most part of the 20th century the research methods have become qualitative (Gomez and Jones III, 2010, p. 1), this being imputed to the use of new technologies, especially of satellites, which began to be used by geographers in almost every branch of the subject. The basis for the theories which substantiate the analysis and method types is, according to the authors, an ontological and epistemological one, from where all scientific paradigms result. In a diagram that they present as a theoretical framework, the two determine dual relations between the elements composing it, exactly because the reason for "that arrows work in both directions indicates that theory needs to be responsive – constantly amended and reworked – in light of the surprises and contradictions that emerge in concrete research activities" (Gomez and Jones III, 2010, p. 2). Thus, as approached in geography, the theories on the world imply an analysis of the human mode of existence, which determines a strong philosophical influence.

Iain Robertson and Penny Richards (2003) also offer a multidisciplinary approach on the cultural landscape, a good study guide which, from our point of view, should not be absent from the library of any person which passionate about this topic. Thus, in the book "Studying Cultural Landscapes", each author offers the methodological side of cultural landscape studies from the perspective of several subjects involved. As the editors themselves state, the book offers a large perspective on some "study methods of symbolic landscapes" (Robertson and Richards, 2003, p. 14).

Therefore, the first chapter of the book refers to the way in which we read the architectural landscape. The approached perspective is not however the one of an architect but of a photographer. By the study he suggests, Iain Black focuses from the very beginning on the historical side of the architectural analysis, the main field of analysis of the author being the relationship between "money, architecture and social power in the financial centre cities of Great Britain and of British Empire between 1750 and 1950" (Black, 2003, p. 19). His approach is a critical one towards the way in which cultural landscape studies use the architecture in subjects such as human and cultural geography. In the third part of the work, the author determines a series of methods of architectural landscape study, keeping the same critical side in his approach. Thus, the author presents the research methods in the same critical manner, making reference especially to the approaches of Mona Domosh and D. W. Holdsworth. The entire argument is built based on the historical architectural analysis and on the thorough study of archives. At the end of the chapter, the author underlines the vital importance that the architecture has within the cultural landscape, emphasizing four important aspects which determine what is called thick description (Black, 2003, p. 28). Briefly, they are as follows: (1) analysis of historical sources and archives; (2) knowledge of the importance that the political and ideological part has within the architectural product and process; (3) knowledge of some iconographic analysis methods in order to underline the importance of the esthetical side of architecture and (4) knowledge of the connection that the architectural product has to the social, cultural and economic sides of life.

The sixth chapter of the book of Iain Robertson makes reference to the way in which the landscape is an engine of the society which produces identity. The author underlines since the very beginning the way in which she relates to the landscape, which has a representational role, conferring uniqueness to it, either it is the real one or only artistic representations of the latter. The discourse on the way in which identities is critical one, presenting different aspects of the identity process. One of the methods proposed for understanding the identity regarding the role of the landscape, is taken from William Mitchell, who refers to the way in which the change of "landscape" into a verb would determine much easier the way in which relate to it. This would help us to see it as a "process through which social and subjective identities appear" (Mitchell quoted in Brace, 2003, p. 123). The presentation was a very solid one and strongly motivated, but at the level of methodology, we must only retain the way in which we are incited to see identity as a created product and not as a certain datum. A product of power, as the author says (Brace, 2003, p. 138), "(...) a belief that identity formation can be thought of as a narrative process in which landscape play a part" (Brace, 2003, p. 138).

At the same time, regarding the importance the rural has in cultural landscape studies one can read in almost all approaches of this topic. Keith Halfacree connects the rural exactly to the idea of identity that Cathrine Brace analyses, arguing the fact that "the rural landscape holds a special place within many cultures" (Halfacree, 2003, p. 142). The author also makes reference to the studies of Denis Cosgrove and Stephen Daniels who influenced by their theories the way in which geographers relate to the landscape, granting an equal importance to its cultural and symbolic side, as well as the material one. The argument emphasizes the way in which Stephen Daniels referred to a landscape duplicity of which is also transferred to the rural level.

Richard Schein, a geography professor at University of Kentucky, USA, specialized in the cultural landscape study, presents on the site of the university the interest it has in the study of this concept and the fact that his works appear in some of the most important papers in the field can only justify the authority he earned in the field of cultural landscape study. When presenting his methodology, he determines a study framework and the way in which he understands the term "cultural landscape". In our thesis we have also mentioned a part of these theories and concepts, such as the cultural landscape as palimpsest. Thus, the author states that "the cultural landscape is a geographical palimpsest, or an accumulation of <<geo-graphy>> - literally human <<earth writing>>. It is a material record of our activity, and as such we can gather information on its creation and meaning, through many sources, but especially through historical records collected in archives" (Schein, 2010, pp. 222-223). This characteristic of being a palimpsest is also emphasized by the fact that in the structure of the cultural landscape is equally integrated to some extent the spatial and visual structure of the way in which the world we live in is organized at the social, cultural, economic level, which determines its ambiguity, as the author admits himself (Schein, 2010, p. 223). This is also given by the way in which the concept is understood and used. Thus, Richard Schein mentions the way in which the term "cultural landscape" is used in different subjects (referring here to the contribution of Paul Groth and Chris Wilson to the transdisciplinary treating of the concept). In order to find a solution, the advice of the author is that when different subjects deal with the same concept but uses different meanings of the latter, "the trick when coming up against this potential communicative obstacle is not to withdraw into the parochialism of your own discipline, or your own pedantic definitions of landscapes or interpretative and analytical preferences, but to search for <<common ground>>, in realizing that we all are interested in some aspect of the tangible, visible scene" (Schein, 2010, p.223).

At the same time, in completion of what we mentioned above, the author states the fact that the cultural landscape is indeed something complicated but at the same time fascinating (Schein, 2010, p. 224), offering us the possibility to study it at several levels from the material on to the symbolic one, passing through a set of multiple methods. Among these, we can mention the visual and the semiological one and psychoanalysis. Richard Schein also appeals to study methodologies offered by Gillian Rose in the book Visual Methodologies, in which is developed the script according to which we can analyze a certain scene only by its visual structure, such as an image. This idea is also transferred to the study of geography, "(...) assuming that landscapes simply <<speak for themselves>> and reminds us that cultural landscapes are themselves representations embedded in, and that embed meaning in, everyday life" (Schein, 2010, p. 225). But this is not enough. It is however necessary, the visual side being important, regardless of the fact that it is either the direct study or the study related to archives, to documents or to scientific volumes. The main sense is generally seen in any reading process, process which supposes understanding, information filtration by every individual from the perspective of his cultural experience and existence.

Thus, we have to remind one of the legitimate questions formulated by Richard Schein who tries to understand exactly the reason for which we analyze cultural landscapes (Schein, 2010, p. 226). Hence, it seems that the question is not a rhetorical one, but certainly is a difficult one to answer. In the idea that a certain answer exists, Schein determines four categories in which the cultural landscape could be fitted (Schein, 2010, p.226). Thus, the first category makes reference to a history of landscape as regards the way in which a certain landscape was created, in which it appeared etc. The second category refers to the significance of landscape. In the first place, it would refer to its significance for the individual or collective identity and in the second place to the way in which landscape is interpreted and the way in which it receives a certain meaning. The third category refers to the landscape as materialized discourse. By giving these categories, for the author one thing is sure, namely the fact that there will always be landscape studies which will never be fitted in none of these or which could have been fitted either in all the four ones or only in one of them (Schein, 2010, p.226).

The argumentation continues supporting the fact that there is the possibility of finding information on the cultural landscape in many journals, archives, magazines or books, many of them not making a reference to this term. This leads to understanding the fact that cultural landscape studies are difficult ones and involve almost always a multidisciplinary team. Richard Schein asserts that "There is an enormous amount of information about cultural landscapes out there, especially when we realize that we (as students of the landscape) are not the only ones who are interested in landscape. (…) The trick is to figure out who else is interested in <<the cultural landscape>> even as they might not think of cultural landscapes per se" (Schein, 2010, p. 227). For the cultural landscape research, its approach from the perspective of a palimpsest is easier, since it helps us understand the concept we operate with as a collection of human interventions over the terrestrial surface. Each of these interventions is nothing but a phase from a continuous transformation process. In the sense of data collection for such a study, Richard Schein gives a series of very useful advices including various methods, from the research in archives to direct observation, interviews with the inhabitants, reading the specialty literature, various tourist guides or even literary works of local authors, as well as the analysis of satellite images, of territory management plans, of local urban planning documents, of historical plans, of aerial photos or of historical ones compared to the contemporary ones (Schein, 2010, pp. 228-237).

Another question as legitimate as the first one that we mentioned refers to the way in which we use the data gathered on the cultural landscape. Richard Schein states that "the harder part is figuring out what to do with the information you collect once you have it, for the most important lesson to remember when collecting <<data>> about a landscape is that the data and the landscape do not speak for themselves" (Schein, 2010, p. 237). To continue this affirmation, the author underlines the fact that even the simplest landscape determines the need for issuing a point of view, an interpretation or a presentation of a certain perception. Thus, "the best landscape descriptions and interpretations and analyses require a basic understanding of the empirical particulars of a cultural landscape in order to ask questions about what that landscape does, about why it is important, about how people live in and through that particular landscape and to what consequence" (Schein, 2010, p.237).

At the end of the presentation of research methods in the cultural landscape study, from the perspective of geography, the author concludes however that any of the research methods specific to human sciences is appropriate for the cultural landscape study, this because "(...)cultural landscape is seen as an important component of everyday life, a central aspect of human existence that is subject to the same sorts of methodological considerations and possibilities as any other human geography problem, issue, hypothesis, or description" (Schein, 2010, p.237).

8.2 Conclusions on the role of the theory within the cultural landscape study

Even if it does not cover extensively this side of research methods in the cultural landscape study, the specialty literature offers however a very wide range of studies on this topic. Two titles that we should mention have been however the object of our research in an entire chapter dedicated to the emphasis of some axioms in defining the concept "landscape" by means of the cultural paradigm. Her we refer to the books "The interpretation of Ordinary Landscapes" edited by D. W. Meinig and "Understanding Ordinary Landscapes" edited by Paul Groth and Todd W. Bressi. These two volumes offer a series of scientific works which represented for a long time and still do a milestone for cultural landscape studies. They are considered compulsory readings for every individual involved in the cultural landscape study, each of them offering a multidisciplinary approach. For example, we will refer again to Professor Richard Schein, who in his seminar on cultural landscape at University of Kentucky offers an increased importance to these titles among others. In the seminar presentation[80] found on the site of the above mentioned university, Richard Schein states that "underlying the seminar is the conceptual premise that the

[80] The seminar syllabus can be found at the address www.uky.edu/AS/Geography/People/Faculty/Schein/Schein_Geo711.pdf. This is the version from the spring of 2005 which totals four pages, thus we will refer to it as follows: (Schein, 2005, p. page no.).

cultural landscape is everywhere and always implicated in the never-finished processes of cultural and social reproduction; or, simply put, that landscape is not innocent" (Schein, 2005, p. 1). Therefore, the purpose of the study is to explain the concept "cultural landscape", to facilitate its understanding by reading the most important theoretical works and through the accumulated knowledge, students should be able to elaborate a coherent and at the same critical and solidly motivated study.

Another example of title known in the field of cultural landscape study is the book The cultural landscape. An introduction to Human Geography of James Rubenstein, professor at Miami University, Oxford, Ohio. It offers an exhaustive study of human geography and of what means the cultural landscape for this subject. The book is divided into 14 chapters, each of them dedicated to a category of cultural landscape. Among others, we should remind the ones regarding the population, migration folklore and popular culture, language, ethnicity, agriculture or urban form. A study guide elaborated by Marc Healy, professor at Elgin Community College, is dedicated to this volume totaling five hundred pages. Reading these two volumes will offer to anyone who wished to accumulate knowledge the wide framework of what cultural landscape means, the concept permeability, due exactly to the numerous elements forming its structure, as well as the difficulty of realizing an ample study on the latter.

Many other titles may be added to the above mentioned ones. But the purpose of our argumentation is not only to enumerate a series of titles of some recognized works but to point out the importance that the theory which formed around the concept "cultural landscape" has in the understanding of the way of operating with the latter and to be able to contribute to its development. This matter becomes all the more important as the term is used more and more in the political, civic or mass-media environment, superseding the borders of the academic environments. But even in the latter, the term ended up by being used in different subjects and the knowledge of theory which is at the basis of its evolution is often ignored. In this regard, Phil Hubbard stated that "the majority of undergraduate geography students have an aversion to study theory and do not understand or appreciate its importance" (Hubbard et. al., 2005, p. viii). This happens even if "put simply, a theory is a set of ideas about how the world works. As such, it is the means by which geographers seek to describe, explain or predict aspects of the world" (Hubbard et. al., 2005, p. 3).

Nevertheless, the truth about the way in which we relate to theory both in geography and in other human sciences among which architecture and urbanism, is simply presented by Phil Hubbard. He thinks that "while it is tempting to dismiss theory, or to try to avoid it because it seems difficult, the reality of doing geography is that we cannot avoid theory. In short, theory infuses the practices of academic geography" (Hubbard et. al., 2005, p. 4). This statement is also valid in case of architecture and urbanism. Especially in case of the latter which has a more tight connection to the practices of the geography colleagues and the sensitivity to space knowledge, social practices, life of individuals, everyday life is more accentuated in our profession. As such, the thorough knowledge of theories when we speak about or we operate with the term "cultural landscape" is necessary, all the more as "there is an important link between theory and praxis. Praxis concerns how theoretical ideas are translated into practice through research, teaching, discussion and debate" (Hubbard et. al., 2005, p.4).

But as researchers in the field, we have an advantage, which is given by reflection and desire for development. Moreover, as architects and urban planners, we are part of a "set of studies connected by their common subject. (…) All these studies refer to the same important thing: humankind – which they describe, narrate and judge and about which they elaborate concepts and theories" (Dilthey, 2001, p. 33). In the field of research, applicable both to our profession and to geography, according to Wilhelm Dilthey (2001, p. 34) two tendencies break through. The first tendency refers to the way in which we relate to and study nature, due especially to the fact that "we are able to control physical world by studying its laws. They can be discovered only if the way in which we feel nature, our involvement in it and the feeling with which we enjoy it surpass the abstract understanding of the world by notions such as the space, time, mass and movement. All these factors combine in order to make the man to place himself on the second plan so that his impression should be able to draft this huge object, the nature, as a structure governed by laws. Thus, it becomes the centre of reality for the human being" (Dilthey, 2001, p. 34). The second tendency, according to the same author, "is the return to experience, the only way by which we have access to nature and life, the only source of significance, value and purpose" (Dilthey, 2001, p.34). And the basis of research and knowledge is represented by these very tendencies which determine "the processes by which meanings are built, negotiated and experienced" (Jackson, 2003, p. 180).

Therefore, in the line of what has been asserted in this paper, the attention must be focused on the knowledge of theories and methods of theoretical study on the concept "cultural landscape" so that when we talk about it or write about it or analyse it, we should be totally aware of what it means, what it was and what it has become, of the ideological and structural changes which made it take the form that we meet so often nowadays. However, stopping only on one of its meanings, as we frequently find in our everyday life, believing that this concept may be understood only by means of a single definition, it would an imprudence of us as professionals. For instance, we should not forget that when we relate to books and studies "they always contain a connection between what exists outside and is perceived by senses and something we cannot reach which is inside" (Dilthey, 2001, p. 34). This also happens with the cultural landscape study, which cannot be studied only by the analysis of some texts or by direct analysis through senses. There should always be a clear connection between the two methods. But as we will notice in the suggested paper, the direct analysis is more and more ignored since we already have at our disposal the instruments (tools) specific to our profession which facilitate an easy study by means of technology and virtual space.

We can thus conclude that in order to reach an understanding of the cultural landscape we have to perform a solid research and to possess the will to progress and to reflect on both meanings of the term. And as a conclusion, we refer to the argumentation of Jeffrey Alexander who stated the fact that "discovering the nature of internal and subjective structures means discovering their significance. The significance is accessible to analysts only of they interpret the very cultural studies and do not try firstly to explain them as results of some other forces, this being the methodological impulse of mechanical approach. The interpretation comes from the understanding, not only from the observation" (Alexander, 2001, p. 7).

9 Contributions to the development of new theories in the theoretical study of the concept of cultural landscape

9.1 The concept of Weltanschauung and the role of human experience in the cultural landscape studies

As one will notice in this chapter, the phenomenological approach of Tilley on the landscape, the metaphorical approach of Italo Calvino through the dialogue between Marco Polo and Kublai Kahn, together with the experience of a landscape language and semantics, interpreted based on the experience and the process of space recognition, together with the creation of situations at the level of knowledge of the urban space, all of these are part of what Martin Heidegger called Weltanschauung (Heidegger, 2006, p. 26)[81], namely a vision of the world (Heidegger, 2006, pp. 26-36). This vision of the world[82], in the opinion of the German philosopher, "refers every time to the everyday life" (Heidegger, 2006, p. 28) and is at the basis of the understanding of the human existence, of spatial and temporal relations (as we will show in the chapter dedicated to cultural landscape chronotopia) and of the primary condition of being (Heidegger, 2006, pp. 27-28)[83]. The topic has also been debated by Mircea Eliade who presents two ways of existence in the world, but we will return to this aspect in our paper.

The cultural landscape study represents, in our acceptance, the very knowledge of human experience, will and behaviour, in the anthropized and/or anthropic space. Translated *ad litteram*, Weltanschauung represents a vision of the world. In this acceptance, the cultural landscape does not propose a single vision of the world, but totals a series of different visions, which by an authentic dialec-

---

[81] Original quotation: "The expression Weltanschauung is a German specific creation, which took place within the philosophy. It appears for the first time in Critique of Judgment of Kant where it has a normal significance: vision of the world in the meaning of world contemplation given by senses or, as Kant said, of mundus sensibilis, thus a vision of world as a simple perception of nature in the widest possible meaning".

[82] The term "Weltanschauung" is also defined on the website http://www.thefreedictionary.com/Weltanschauung, as being "a comprehensive view or personal philosophy of human life and the universe".

[83] Original quotation: "The vision of the world always includes in itself the vision of life. The vision of the world is the product of a global reflection of the world and the human existence, phenomenon which in its turn takes place in various ways – for certain people explicitly and knowingly, for others by overtaking a dominant vision of the world. Every human being grows inside of such a vision and lives in its element. The vision of the world is determined by the environment: people, race, status, cultural evolution phase".

tic, language and narrations, determines the framework necessary to understand the complexity of relations which are at the basis of human action in the anthropic space.

In the approach of the subject regarding the spatial existence of the individual, "to experience is to learn; it means acting on the given and creating out of the given. The given cannot be known in itself. What can be known is a reality that is a construct of experience, a creation of feeling and thought" (Tuan, 1977, p. 9). Thus, appears the need for the existence of a reference element compared to which the surrounding realities could be defined. The duality of existence[84] and the abundance of stereotypes such as culture and nature, beautiful and ugly, mind and body or rational and irrational, lead to the division of the world into what we or I and the other or the others are. The locality leads to relations of the type here and there, close and far, leading to hierarchies and classifications, divisions and delimitations. The existential dualism and the space perception by means of the experience, as well as the definition of the place according to oneself and the other are the very stimuli of identity[85] (Jordan, 2011, pp. 12-13). Thus, the reality that determines the human experience is different according to the position that the individual approaches. One can easily deduct that the experience may have a participative characteristic and a purely observative one. This distinction derives from the very position that the individual has in his initiated path: spectator or actor. For an extended understanding, we refer again to the writings of Yi-Fu Tuan who stated that the experience is tightly related to senses and to the way in which space is perceived (Tuan, 1977, p. 9)[86]. The space is organized according to distances approachable and perceptible by the man and to the type of inter-human relations it operates. Space perception and the relation to it changes with the receiver's age and Yi-Fu Tuan presents very well the way in which with the ageing the space awareness is more and more clear which also influences directly the experience lived by the individual in the respective space. At the same time, the space awareness also implies its more precise locality which facilitates the understanding and the interpretation.

As we reminded in the previous paragraphs, the experience means learning. Therefore, Donald W. Meinig said that "any landscape is composed not only of what lies before our eyes but what lies within our heads" (Meinig, 1979, p. 34).

---

[84] We talk about a dual existence thus making reference to the two parts, the material one and the spiritual one.

[85] Original quotation: "Social groups have usually a stable location in space, they reside more or less permanently in certain places, transform them and leave a certain cultural footprint, which is partly visible and termed a 'cultural landscape' by geographers. (...) Place names mark space-related identity on the one hand and support ties to places on the other. (...) another characteristic of social groups (and individuals) relevant in our context is that they discern between 'self' and 'the other', between what is theirs and what is the other's. This attitude is unavoidable to preserve (group) identity, although it may look anachronistic in our globalizing world".

[86] Original quotation: "Experience is directed to the external world. Seeing and thinking clearly reach out beyond the self. Feeling is more ambiguous".

Yi-Fu Tuan presented the experience as a series of events which marked the life of an individual, a standard of his maturity (Tuan, 1977, p. 9)[87]. At the same time, he refers to a sensorial experience as way of understanding and identifying the space. Contrary to this passive vision of the experience, Elizabeth Ellsworth presents its participation side in which the man is the very experience he lives[88] (Ellsworth, 2005, p. 26), being directly involved and not at all passive as in the vision of Yi-Fu Tuan.

In the same order of ideas, Brian Massumi states that from the phenomenological point of view, the terminus point of the experience is exactly the individual. The personal side pre-existed in the world, from the perspective of the phenomenology and through the process of perception is reflected what already exists and is preset[89] (Massumi, 2002, p. 191). Thus, Elizabeth Ellsworth argues that every object or event has an educational role and the architectural space or the natural ones are real learning environments[90] (Ellsworth, 2005, p. 38). However, the everyday rhythm, by the speed with which it takes place, lives no time for the necessary break in order to perceive and understand the environment as a constant "teacher" in our lives. We live the moments in which the hurry is so accentuated that what surrounds us has no value anymore and attracts no interest. Only as tourists we dedicate the time necessary to the contemplation and valorization of the potential offered by the environment. Space creators, those who shaped the architectural and urban space are real pedagogues of this world (Ellsworth, 2005, p. 38)[91].

The topic of the spatial experience is also present at Mircea Eliade. According to him, there are two ways of existing in the world, and thus two ways of experience: sacred and profane (Eliade, 2005, p. 15). The author says that the sacred

---

[87] Original quotation: "Experience has a connotation of passivity; the word suggests what a person has undergone or suffered. (...) To experience in the active sense requires that one venture forth into the unfamiliar and experiment with the elusive and the uncertain".

[88] Original quotation: "In other words, experience is not subjective in the sense that "I have" an experience. When we think of experience as a question of sensation, we remind ourselves that we do not have experiences. We are experiences. The "I" of an experience does not precede the experience".

[89] Original quotation: "That the personal is the finale distinguishes this synesthetic ontogenesis of experience from phenomenological approaches. For phenomenology, the personal is prefigured or "prereflected" in the world, in a closed loop of "intentionality." The act of perception or cognition is a reflection of what is already "pre-" embedded in the world. It repeats the same structures, expressing where you already were. Every phenomenological event is like returning home".

[90] Original quotation: "Architecture becomes pedagogical, pedagogy becomes architectural when together they create a fluid, moving pivot place that puts inside and outside, self and other, personal and social into relation. A performance becomes pedagogical, pedagogy becomes theatre when together they create the potential to disrupt and refigure both inside and outside — both self and society".

[91] Original quotation: "At the hands of these designers, pedagogy becomes a dynamic that creates the experience of an idea, of a way of making sense of self, the world, and self in the world. It becomes the force that sets interior self-experience in motion to encounter the outside "not me." Pedagogy stages encounters with the unthought — encounters with the future as in the making".

space is the real one for the religious man and the experience of the sacred space has an existential value (Eliade, 2005, p.20). In the author's opinion, the profane space is a homogenous and neutral space and the profane experience is a purified experience of the individual who rejects the sacrality (Eliade, 2005, p.20), even if "in this experience of the profane space values keep intervening which remind more or less of the lack of homogeneity which characterizes the religious experience of the space. There are privileged places, qualitatively different from the other ones: the region of birth, the place of the first love, a street or a corner from the first foreign city seen in the youth. All these places keep, even for the most unreligious man, an exceptional <<unique>> quality because it represents <<holly places>> of his private Universe as if this religious being would have had the revelation of a reality different from the one in which he takes part by his daily existence" (Eliade, 2005, p. 21).

Our interest towards the presentation of Mircea Eliade as regards the space experience consists of the way of the place recognition or, as the author himself asserts, at the "repetition of Cosmogony" (Eliade, 2005, p. 27): "the installation in a territory means in fact its recognition. (...) The installation in a certain place, its organization and dwelling are actions which imply an existential choice: the choice of a Universe that will be assumed by creation" (Eliade, 2005, p. 29).

9.2 Locality and recognition of space in the cultural landscape study

In the line of the aspects stated in the previous chapter, Christopher Tilley argued[92] (Tilley, 1994, p. 10) that it is not all about a single space but about several spaces which are results of the human activity, of the impact of various cultures of different peoples. Thus, the spatio-temporal compression determined by the change of the human society leads to a different interpretation of the cultural landscape. This aspect determines an important characteristic within the theory of the concept which implies the existence of a guide in the interpretation and reading process of the cultural process.

However, before starting the analysis of the interpretation of the cultural landscape by means of the processes of locality and recognition of space, we must provide the definition given by the Longman Dictionary of Contemporary Eng-

[92] Original quotation: "The alternative view starts from regarding space as a medium rather as a container for action, something that is involved in action and cannot be divorced from it. As such, space does not and cannot exist apart from the events and activities within which it is implicated. Space is socially produced, and different societies, groups and individuals act out their lives in different spaces. Space in itself no longer becomes a meaningful term. There is no space, only spaces".

lish[93]. For the clarity of the study, we extract the meaning of the word being the one of understanding, transferring the content by appropriate ways, explaining. Accepting these uses of the term, we understand to a certain extent the connection between interpretation and experience (we will return to this term consequently). In order to explain a phenomenon, a landscape, an action, to be able to render its meaning, the individual must understand that phenomenon, landscape or action. Thus, as one will be able to subsequently distinguish in this paper, Peirce Lewis stated before the presentation of his axioms, the fact that in the reading process of the landscape a guide, such as the grammatical one, was necessary. A guide that should help understanding the cultural landscape. Consequently, his axioms are the result of one's own experience[94] (Lewis, 1979, p. 15) acquired over time.

9.2.1 Understanding the concepts of space and place

Understanding the phenomena which determined cultural landscape changes and adaptations is part of a phenomenology of the latter. It is this thing that is presented by Christopher Tilley, even though its field of study includes anthropology and archaeology. The author offers a useful presentation of what he calls phenomenological perspective[95] of the cultural landscape. According to Tilley's description, the phenomenology implies the understanding and the description of elements from the perspective of the individual (Tilley, 1994, pp. 11-12)[96]. The author refers to the phenomenology of Heidegger and Merleau-Ponty in order to create the framework necessary for the study of the landscape phenomenology. The investigation of the concept of dwelling suggested by Heidegger opens for Tilley a new perspective of what means the relation be-

---

[93] From the online version of Longman Dictionary of Contemporary English, we extract the following definitions: "the way in which someone explains or understands an event, information, someone's actions etc"; "the way in which someone performs a play, a piece of music etc and shows what they think and feel about it".

[94] Original quotation: "What we needed, I concluded, were some guides to help us read the landscape, just as the rules of grammar sometimes help guide us through some particularly convoluted bit of syntax. Little by little, I began to write down some of the rules that I discovered over the years of looking and learning and teaching about American landscapes and which I found helped me understand what I saw. I call these rules "axioms", because they now seem basic and self-evident, as any proper axiom might be".

[95] According to the Merriam-Webster Dictionary, the term is defined as follows:
1: the study of the development of human consciousness and self-awareness as a preface to or a part of philosophy. 2. a(1) : a philosophical movement that describes the formal structure of the objects of awareness and of awareness itself in abstraction from any claims concerning existence (2): the typological classification of a class of phenomena <the phenomenology of religion>
b: an analysis produced by phenomenological investigation.

[96] Original quotation: "The key issue in any phenomenological approach is the manner in which people experience and understand the world. Phenomenology involves the understanding and description of things as they are experienced by a subject".

tween individual and landscape and concepts of space and place (Tilley, 1994, p. 14)[97].

The theoretical discourse connected to the place and space has always had a specific status in the specialty literature, being an important research subject for geography in general by also for other subjects such as the anthropology, archaeology or philosophy as well as architecture and urbanism or territory management. Constantly enriched with new approaches, the discourse is not yet completed. Although sometimes it can be extremely abstract and difficult to accept even by specialists, the understanding of the concept "place" is tightly connected to conceptual spatial delimitation (Tilley, 1994, p. 15)[98]. Thus, Tilley listed five delimitations of the space (Tilley, 1994, p.15) that we will present hereinafter. Briefly, he talks about the somatic space, the perceptual space, the existential space, the architectural space and the cognitive space. In two pages, the author presents the five above mentioned types of space, thus fitting in terms of concepts the space and place study (Tilley, 1994, pp. 16-17). Thus, the somatic space is defined by the author as the space of sensorial experiences and of the movement of the human body (Tilley, 1994, p. 16)[99]. The perceptual space is defined by Tilley as the "egocentric" space, the space that defines human experiences, the memory of the place associated to the experiences, feelings and needs of the individual when he is in a certain space (Tilley, 1994, p.16)[100]. The existential space is assimilated as the basis on which the percep-

---

[97] Original quotation: "While 'dwelling' occurs in different varieties and textures of humanly created space, this social existence is, of course, rooted in natural and non-humanly created environments. The concept of dwelling, with its fourfold ontological implications as pointed out by Heidegger, the human body as a focus for the perception of a humanized world, and the groundedness of social Being in that which is not humanly created constitute the fundamental presuppositions for beginning to think about the relationship between people and landscape in a fresh manner. Subjectivity and objectivity connect in a dialectic producing a place for Being in which the topography and physiography of the land and thought remain distinct but play into each other as an 'intelligible landscape', a spatialization of Being".

[98] Original quotation: "Without places there can be no spaces, and the former have primary ontological significance as centres of bodily activity, human significance and emotional attachment. The meaning of place is grounded in existential or lived consciousness of it. It follows that the limits of place are grounded in the limits of human consciousness. Places are as diffuse and differentiated as the range of identities and significances accorded to them. (...) There may be a strong affection for place (topophilia) or aversion (topophobia), but places are always far more than points or locations, because they have distinctive meaning and values for persons".

[99] Original quotation: "An understanding of this space takes as its starting-point the upright human body looking out on the world. Space opens out before the body and is differentiable in terms of front/back; left/right; vertical/horizontal; top/bottom; within reach/beyond reach; within hearing/beyond hearing; within sight/ beyond sight; here/there polarities".

[100] Original quotation: "Perceptual space is the egocentric space perceived and encountered by individuals in their daily practices. The centre of such a space is grounded in individual perception of distances and directions, natural objects and cultural creations . (...) A perceptual space is one that links patterns of individual intentionality to bodily movement and perception. It is a space of personality, of encounter and emotional attachment. It is the constructed life-space of the individual, involving feelings and memories giving rise to a sense of awe, emotion, wonder or anguish in spatial encounters. Such a space may as often as not be felt rather than verbalized.

tual space appears, the space in which the entire human activity takes place, the real space, of physical existence (Tilley, 1994, p.16)[101], of social condition of the individual. Therefore, the architectural space is the bridge between the existential space and the perceptual one. The architectural space is the one defining the two above mentioned spaces, the existential and the perceptual ones (Tilley, 1994, p. 17)[102]. Consequently, the cognitive space is the space of analysis and reflection (Tilley, 1994, p.17)[103].

Christopher Tilley defines very well the way in which the architectural space is the one that creates spatial relations, the one that defines the interior or the exterior. However, we can make an addition, noticing the rather slightly incomplete use of this term. We must clearly distinguish between the architectural space and the urban space under the circumstances in order to have a uniform and pertinent image on these elements. Tilley was not wrong as he reminded and defined the type of space as being architectural. Essentially, the urban space is a complex system consisting of several architectural spaces, each of them having a well defined set of relations which determines the individual's existence in the built environment. Both the urban space and the architectural space are spaces of a built reality based on ideologies and doctrines, on visionary ideas. They are spaces of power based exactly on segregation, distinction, control, dominance[104].

It creates personal significances for an individual in his or her bodily routines - places remembered and places of affective importance".

[101] Original quotation: "Perceptual space is intricately interlinked with existential space or the lived space as it is constructed in the concrete experiences of individuals socialized within a group. The meanings of existential space transcend the individual and form a grounding for perceptual space rather than being some kind of summation of individualized perception. Existential space is in a constant process of production and reproduction through the movements and activities of members of a group. It is a mobile rather than a passive space for experience".

[102] Original quotation: "Architectural space only makes sense in relation to pragmatic, perceptual and existential space, but involves a deliberate attempt to create and bound space, create an inside, an outside, a way around, a channel for movement. Architecture is the deliberate creation of space made tangible, visible and sensible. This is why buildings play a fundamental role in the creation and recreation, production and reproduction of existential space and have profound structuring effects on perceptual space".

[103] Original quotation: "Finally, cognitive space provides a basis for reflection and theorization with regard to understanding the others. It is the 'space' of this discussion and analysis. Space can only exist as a set of relations between things and places. In this sense there is no space that is not relational. Space is created by social relations, natural and cultural objects. It is a production, an achievement, rather than an autonomous reality in which things or people are located or 'found'. Having been constituted by things and places spatial relations affect the way in which they relate. In other words there is a sociospatial dialectic at work – space is both constituted and constitutive".

[104] We can hereby remind the film Baaria made in 2009 by the Italian producer Giuseppe Tornatore, an autobiographical film which was shown at the opening of the Venice Film Festival in the same year and which describes, with the characteristic sarcasm met in the Italian comedies, a blind urban consultant. The message is subtly and elegantly shaped, presenting the vices of a sick and corrupt society (the officer approved the urban plans only after he had received the money).

The architectural space (co)exists with the urban space and offers to the latter consistency, form, coherence. The relations between architectural spaces, combined with the perception space and to the existence one, define the urban space itself. The city is a complex system in which the different types of spaces listed by Tilley can be found. All these five types exist in what we call urban space. The urban space is a meta-space of relations between the other types of spaces. However, all of these also interact with another type of space which in its turn also covers the urban space. This is the large space of landscape. We will return to this matter in the following paragraphs. We will hereinafter consolidate the discussion on the spaces listed by Tilley, especially on that of the omitted space, the urban one. The topic of the individual's urban existence, as well as of the way of relating to his existential space, has been used frequently in the urban theory of the 20th century. In our opinion, the urban space is the very space of the human existence in the complex system of relations that we call city, and the architectural space is the intimate space, of the individual's private existence that can be represented by his house, the yard of the house, the working space or the fun space. Kevin Lynch suggested a series of questions connected to the form that human settlements must have and which are the principles that we follow when elaborating these forms. What is important in the answer that the author himself gives is the way in which he combines human actions in the urban space with information and good flows, with the existence of architectural constructions and with the cyclicity of nature (Lynch, 1984, p. 48)[105]. The city evolved over time and the communities ended up by being cells of a central system based on expression of power and control. From the first religious communities to the guilds, to the bourgeoisie and then to capitalist corporations, without forgetting about monarchies and governs, each of them has contributed to the production in the urban space of power and control symbols on various low social classes. The perceptual space and the one of the human existence ended up by being spaces defined by doctrines and practices of some human societies with varied cultures and identity values, but which essentially share the same needs and answer to the same requirements of

---

This symbolizes the malignous practices of some past times which still exist, being present in certain modern societies.

[105] Original quotation: "The fundamental problem is to decide what the form of a human settlement consists of: solely the inert physical things? Or the living organisms too? The actions people engage in? The social structure? The economic system? The ecological system? The control of the space and its meaning? The way it presents itself to the senses? Its daily and seasonal rhythms? Its secular changes? Like any important phenomenon, the city extends out into every other phenomenon, and the choice of where to make the cut is not an easy one. I will take the view that settlement form is the spatial arrangement of persons doing things, the resulting spatial flows of persons, goods and information, and the physical features which modify space in some way significant to those actions, including enclosures, surfaces, channels, ambiences, and objects. Further, the description must include the cyclical and secular changes in those spatial distributions, the control of space, and the perception of it. The last two of course, are raids into the domains of social institutions and of mental life".

the urban human condition (Tilley, 1994, p. 17)[106]. In order to better understand this, we will remind Italo Calvino who in his book Invisible Cities presented in the form of poems a conversation between Marco Polo and Kublai Kahn in which he explored the imaginative side of cities, presenting over fifty-five of them. We will refer to what stated Kevin Lynch on the book of Italo Calvino at the end of the third chapter of the book Good City Form (Lynch, 1984, p. 72)[107]. He briefly presents the descriptions of cities offered by Marco Polo, reasserting the actuality of relations between individuals and inhabited spaces. This actuality is also connected to the way in which we perceive the space (Calvino, 1972, p. 137)[108], the landscape, the territory, the city and, in his conversation, Marco Polo clearly points out the fact that regard-less of the person telling the story (we refer here to the landscape, to urban space, to the territory as a metaphorical narrator), the interpretation belongs to the listener (namely to the spectator, to the individual directly involved in the urban life, to the space explorer) (Calvino, 1972, p. 135)[109]. We can conclude the fact that, in his existence, the individual defines his human and social condition by relating to a series of interconnected spaces. It is the very existence of the latter that leads to large spaces by the process of locality and appropriation of a space, to shaping the concept and the entity of place. The presence of the man in the landscape, the need for the adaptation to the surrounding nature, the need for the understanding of processes and of elements composing the landscape, as well as the inevitable evolution of the human race against the other living creatures have led to the space particularization, to its adaptation to the needs of an evolved social existence.

---

[106] Original quotation: "Places may, of course, be experienced and conceptualized at any number of spatial levels, from personal space to community space to regional space and so on. Places overlap according to scales of action, interest, movement and concern. Place is an irreducible part of human experience, a person is 'in place' just as much as she or he is 'in culture'. Place is about situatedness in relation to identity and action".

[107] Original quotation: "Each city is a society that exaggerates the essence of some human question, and for each there is a form, brilliantly and surprisingly conceived, that fulfills and informs that question. Polo talks of desire and memory; of diversity and routine; of the temporary and the permanent, the dead, the living, and the unborn; of images, symbols and maps; of identity, ambiguity, reflections, the seen and the unseen; of harmony and discord; of justice and injustice; of mazes, traps, and endlessness; beauty and ugliness; metamorphosis, destruction, renewal, continuity, possibility, and change. The dialog is a great panorama of utopia and cacotopia, which explores, in a wonderful, circling fantasy, the relations between people and their places".

[108] Original quotation: "Traveling, you realize that differences are lost: each city takes to resembling all cities, places exchange their form, order, distances, a shapeless dust cloud invades the continents. Your atlas preserves the differences intact: that assortment of qualities which are like the letters in a name".

[109] Original quotation: "Kublai asks Marco, 'When you return to the West, will you repeat to your people the same tales you tell me?' 'I speak and speak,' Marco says, 'but the listener retains only the words he is expecting. The description of the world to which you lend a benevolent ear is one thing; the description that will go the rounds of the groups of stevedores and gondoliers in the street outside my house the day of my return is another; and yet another, that which I might dictate late in life, if I were taken prisoner by Genoese pirates and put in irons in the same cell with a writer of adventure stories. It is not the voice that commands the story: it is the ear".

9.2.2 The locality process of space

The locality supposes the acceptance of the "context of functional vocations" which has been mentioned by Heidegger (Heidegger, 2006, p. 270). The locality means (de)limitation and enclosure and is determined both in space and in time[110]. The space and place thus have an important place in defining the social relations and the way of human interaction and are found at the basis of the individual's cultural identity. The locality also contributes to the determination of a possible cultural meaning attributed to the landscape. Once it has been named and located, the space becomes a place and acquires significance. It is transformed according to the preferences of the individual and to his experience (Tuan, 1977, pp. 9-17)[111].

In order to exemplify, we would like to approach hereinafter the frequent problem found in the specialty works, from geography to architecture or from philosophy to history: the one of the space awareness and understanding, namely of giving meaning and significance to it.

It is exactly the definition provided by Christian Norberg-Schulz to the term of place that we propose for the exemplification. He takes into consideration the numerous events that define the daily life in order to be able to determine what place means. For Norberg-Schulz, the place is more than an abstract locality (Norberg-Schulz, 1979, p. 6)[112], is simply a part of the human existence (Norb-

---

[110] According to the Merriam-Webster online, the definition of the term is as follows: "plural lo•cal•i•ties. Definition of LOCALITY 1: the fact or condition of having a location in space or time; 2: a particular place, situation, or location. First Known Use of LOCALITY:1628". Site accessed on May 11, 2011, 12:41 AM.

[111] Yi-Fu Tuan makes a subtle and pertinent description of the transformation of the space into place and of the importance the experience has when defining the place. "Experience is a cover-all term for the various modes through which a person knows and constructs a reality. These modes range from the more direct and passive senses of smell, taste, and touch, to active visual perception and the indirect mode of symbolization. (...) Experience is compounded of feeling and thought. Human feeling is not a succession of discrete sensations; rather memory and anticipation are able to wield sensory impacts into a shifting stream of experience so that we may speak of a life of feeling as we do of a life of thought. (...) Human beings not only discern geometric patterns in nature and create abstract spaces in the mind, they also try to embody their feelings, images, and thoughts in tangible material. The result is sculptural and architectural space, and on a large scale, the planned city. (...) Place is a type of object. Places and objects define space, giving it a geometric personality. Neither the newborn infant nor the man who gains sight after a lifetime of blindness can immediately recognize a geometric shape such as a triangle. The triangle is at first 'space', a blurred image. Recognizing the triangle requires the prior identification of corners – that is, places".

[112] Original quotation: "What, then, do we mean with the word place? Obviously we mean something more than abstract location. We mean a totality made up of concrete things having material substance, shape, texture, colour".

erg-Schulz, 1979, p. 6)[113]. The existence of the place supposes previously the locality of an object or event. One cannot refer to anything of what exists without locating that something.

The specialty literature debating this subject is extremely wide but there are certain aspects that concern us directly in the cultural landscape study. As we have also shown in case of the approach of the cultural landscape as a vision of the real in the book published last year, the study of the phrase sense of place that we find at many authors in the field is very important. Its understanding is also essential in direct relation to the understanding of the term "space" presented in the previous paragraphs.

First of all, an explanation of the term sense is required. According to the Merriam-Webster Dictionary, the term was used for the first time in the 14th century and comes from the Latin word *sensus* which means to feel, to perceive. At the same time, it has an Anglo-French origin from the term *sen* or *sens* which means sensation or mechanism of perception[114].

John Brinkerhoff Jackson associates the above mentioned term to genius loci, considering it a modern iteration of the latter (Jackson, 1994, p. 157)[115]. At the same time, he makes the association of the term to locality that we mentioned above. As the author (Jackson, 1994, p. 157) asserts, in the past the existence of a protective spirit of the place was believed, and the locality was the very process by which a spatial entity was provided at the cognitive and affective level with recognition of its special status due to the presence of a spiritual, protective power. In time, this was transformed into what will represent the influence of the place on the well-being, satisfaction and achievement condition of the individual once he is in that place and the spiritual side was almost completely lost (Jackson, 1994, p. 158).

The analysis of the definition of the term sense also points out the nouns awareness and consciousness. The use of the two terms is known from the influent theories of Yi-Fu Tuan in human and cultural geography. He associates the architectural space to the process of space awareness (Tuan, 1977, p. 102)[116]. Moreover, the act of construction is primordial both for animals and

---

[113] Original quotation: "The concrete things which constitute our given world are interrelated in complex and perhaps contradictory ways. (…) A concrete term for environment is place. It is common usage to say that acts and occurrences take place. In fact it is meaningless to imagine any happening without reference to a locality. Place is evidently an integral part of existence".

[114] The online Merriam-Webster Dictionary presents the origin of the term sense as: "Origin of SENSE: Middle English, from Anglo-French or Latin; Anglo-French sen, sens sensation, feeling, mechanism of perception, meaning, from Latin sensus, from sentire to perceive, feel; perhaps akin to Old High German sinnan to go, strive, Old English sith journey. First Known Use: 14th century". Site accessed on May 11, 2011, 11:38 AM.

[115] Original quotation: "It is a awkward and ambiguous modern translation of the Latin term genius loci".

[116] Original quotation: "A planned city, a monument, or simple dwelling can be a symbol of the cosmos. In the absence of books and formal instruction, architecture is a key to comprehending reality".

for humans and by this process relations are shaped and social order (in case of humans, of course) is imposed. Even through the spiritual side has vanished, in modern societies the religious side exists and within different communities, he church is the symbol of power and the central pole of human interaction in the urban or rural built environment (Jordan, 2009, pp. 34-35)[117]. The central market of the city, of the citadel was the space in which took place the relations between individuals, space which by its symbolic and sentimental significance became a place of processions and meetings. That space acquires a sense, a purpose, a meaning, a conscience, due to the vital role it may have in developing social relations.

The topic is also approached by John Brinckerhoff Jackson who presents the situation in American cities and the way in which the space awareness and the sense granted to the place is not always associated to the architecture or the built space itself, but to social events that happen in a certain space (Jackson, 1994, pp. 159-160)[118]. At the same time, the author underlines that, regardless of the nature of the place in which the individual is, the latter is the one defining the pace and the spiritual status and the emotional significance are important when defining the affiliation to that place. Considering also the historical context different from the Europeans and the urban history in the United States, it is easy to understand the difference between the two approaches regarding the concept of place.

Therefore, the phrase sense of place makes us think of the well-known Romanian proverb *omul sfinteste locul, nu locul pe om* (*homo locum ornet, non ornate homonem locus / it is the man that consecrates the place and not the place that consecrates the man*). Staying in the same field of study, we will bring an understanding of the proverb to our cultural landscape study and we will thus refer again to what Professor Peter Jordan asserted in relation to the place identity. He argues the fact that the individual forms his identity based on the place he lives in, and the cultural landscape is also formed as a consequence of the human intervention (Jordan, 2009, p. 35)[119]. Thus, it is shaped the bipolarity of

---

[117] Original quotation: "In a way, each landscape tells a story about the culture by which it has been shaped. When the historical layers of a landscape are analyzed, a temporal sequence of cultures can also be traced. (…) another striking example in this respect is Cluj-Napoca in Romanian Transylvania [Ardeal]. Although this city is today predominantly Romanian by nationality and Orthodox by domination, the Catholic church of the Hungarians is located right in the centre of the city, at its main square, unmatched by any other landmark".

[118] Original quotation: "As our cities have grown we have come closer together and acquired a more inclusive sense of community. Even so, I'm inclined to believe that the average American still associates a sense of place not so much with architecture or a monument or a designed space as with some event, some daily or weekly or seasonal occurrence which we look forward to or remember and which we share with others, and as a result the event becomes more significant than the place itself. Moreover, I believe that this has always been the common or vernacular way of recognizing the unique quality of the community we live in".

[119] Original quotation: "Apart from this feeling of familiarity and safety the cultural landscape also conveys by this very reason culture. Having been shaped by the culture of a social group (in

the concept "cultural landscape". It is both the engine of shaping the individuals' identities and a scene of their intervention in shaping its own identity. We have used the term "identity" in reference to the cultural landscape in order to facilitate the observation of the dual effect of both entities. Everyone is in its turn both a spectator (passive and active) and an actor/motor of transformations. This relation is governed by a vector of opposed orientation and direction. It is a bidirectional and dual relation. Anne Whiston Spirn stated that the place is a peculiarity of space which bears in it the force of each past experience, being a series of stories old over time. Thus, the place recognition and its sense represent the totality of these stories which are held by every place (Spirn, 1998, p. 160)[120]. The place is a fine fabric of various existential contexts (Spirn, 1998, p. 160)[121].

In the same line, in his studies, Kevin Lynch pays an increased attention to performance dimensions (Lynch, 1981, p. 111) of cities and among the criteria we also find the one of sense. The Romanian word means conscience, wisdom, meaning, purpose, significance or sense. As regards the city and the urban form, Kevin Lynch argues that the meaning of the term sense of settlement (Lynch, 1981, p. 131)[122] as the clarity by which the urban space is perceived and identified. This clarity may be obtained both by the urban form itself and by means of the objective and subjective knowledge of the human factor on it. Completing what Professor Peter Jordan said about the identity, the theoretical way of understanding the phrase sense of place can only be completed by reminding the fact that, in the opinion of Lynch, the identity is the simplest form that the term sense/purpose/meaning can take (Lynch, 1981, p. 131)[123]. The identity, in an extremely brief argumentation, represents the very element

---

practice rather by some more active or respected individuals or by the driving and dominating forces of a group) it reflects this culture back to the group and – this time – to all its members".

120 Original quotation: "Every place has ongoing stories, recognized, concealed, and lost. Some take longer to tell than others, some are short; some have an ending, other are opened, still unfolding. All that has happened in a place, all the lives lived there contribute to a sense of place; identity is defined by the common and by the rare alike".

121 Original quotation: "A place is particular, a tapestry of woven contexts: enduring and ephemeral, local and global, related and unrelated, now and then, past and future. Landscape context is a fabric whose strands are narratives of landscape elements and features, both the persistent and the fleeting".

122 Original quotation: "By the sense of a settlement, I mean the clarity with which it can be perceived and identified, and the ease with which its elements can be linked with other events and places in a coherent mental representation of time and space and that representation can be connected with nonspatial concepts and values. This is the join between the form of the environment and the human processes of perception and cognition. (...) Sense depends on spatial form and quality, but also on the culture, temperament, status, experience, and current purpose of the observer. Thus the sense of a particular place will vary for different observers, just as the ability of a particular person to perceive form varies for different places".

123 Original quotation: "The simplest form of sense is identity, in the narrow meaning of that common term: a sense of place. Identity is the extent to which a person can recognize or recall a place as being distinct from other places – as having a vivid, or unique, or at least a particular, character of its own".

which gives sense to a place, which gives it meaning and significance. As John Brinckerhoff Jackson associated the events as giving birth to place significance and conscience, Kevin Lynch named the way of relating to the space by means of events as sense of occasion (Lynch, 1981, p. 132)[124]. It is basically the same thing, but has different names at the two theoreticians.

We thus have two phrases that, as John Brinckerhoff Jackson reminded, (re)define in fact in a modern manner the well-known genius loci. Simon Bell stated the fact that the term genius loci implies the very existence of an essential characteristic of a place, of a peculiarity which could distinguish it from the others (Bell, 2004, p. 119)[125]. Even so, two identical places do not exist on this earth, just as two identical beings do not exist. Thus, this genius loci is an abstract and ambiguous term (let us remind the aspects mentioned regarding the term of landscape and its ambiguity). Simon Bell reasserts the importance of locality, as well as the individual's identity as belonging to a certain place (Bell, 2004, p. 120)[126].

Christian Norberg-Schulz has determined the structure of the place as being described by the terms "landscape" and "settlement", but the analysis was limited to categories of the type space and character (Norberg-Schulz, 1979, p. 11)[127]. Thus, the way of his argumentation passes by the concepts of space from Giedion and Kevin Lynch to Martin Heidegger, concluding that the essential attributes of a space are the extension and the enclosure (Norberg-Schulz, 1979, p. 12)[128]. The character of a place, stated Norberg-Schulz, is the way in

---

[124] Original quotation: "Events can also have identity; this is the sense of occasion. Special celebrations and great rituals have it in heightened degree. Occasion and place will reinforce each other to create a vivid present".

[125] Original quotation: "The genius loci or the spirit of the place refers to the special, unique quality one place has over another. • Genius loci is intangible but is a highly valued aspect when attached to a landscape, which helps to make it sensitive. • Genius loci can be vulnerable because the factors contributing to it may be difficult to identify. • Relatively extensive yet self-contained areas can possess genius loci as well as smaller-scale places. • Any design must respond to a strong genius loci. • A landscape lacking character can be given a stronger sense of place by good design creatively applied".

[126] Original quotation: "Place is very important to us and in our lives. Our sense of identity may be bound up with a particular place and we may refer to ourselves by this, for example, 'I am a Parisian'. The location itself marks the position of the place, but place itself consists of the totality of the natural and man-made things, assembled in a unique way and may well include the history and associations attached to the place by the people who identify with it. While all places have a character, this in itself is not adequate to induce genius loci. It is the uniqueness which makes it special and with which we can readily associate".

[127] Original quotation:"Our preliminary discussion of the phenomena of place led to the conclusion that the structure of place ought to be described in terms of 'landscape' and 'settlement', and analyzed by means of the categories 'space' and 'character'. Whereas 'space' denotes the three-dimensional organization of the elements which make up a place, 'character' denotes the general 'atmosphere' which is the most comprehensive property of any place".

[128] Original quotation: "The outside-inside relation which is a primary aspect of concrete space, implies that spaces possess a varying degree of extension and enclosure. Whereas landscapes are distinguished by a varied, but basically continuous extension, settlements are enclosed entities. Settlement and landscape therefore have a figure-ground relationship".

which it is "give" (Norberg-Schulz, 1979, p.14)[129] to us and is determined by the formal and material character of the place. In conclusion, the place is the very purpose of the cultural landscape investigation and at the same time the point from which in starts, as Christian Norberg-Schulz himself said (Norberg-Schulz, 1979, p. 18).

9.2.3 Space consecration by the naming process

Therefore, in order to continue the above mentioned aspects, we can argue how the daily existence of the individual led to the appearance of names of places (toponyms, endonyms and exonyms) (Jordan, 2011, p. 10)[130] that gave a sense and a meaning to the cultural landscape (and space). Thus, names created a direct relation by which the man was able to become part of the landscape exactly sue to the way in which he reported to it, to the way in which it described it and contemplated it (Tilley, 1994, p. 18). In a work of the geographers Ion Nicolae and Bogdan Suditu, the toponymy is described as "the unwritten history of a people, a real archive where is kept the memory of so many events, episodes and facts more or less old or important, which took place over the time and impressed to a certain extent the popular soul" (Iordan [1963] quoted in Nicolae and Suditu, 2008, p.10).

Figure 34. An example of consecration by naming. In case of certain localities in Transylvania there are bi- or trilingual plates, according to Law no. 215/2001 chap.1 section 1 art. 17. Copyright: Alexandru Calcatinge.

---

[129] Original quotation: "We have pointed out that different actions demand places with a different character. A dwelling has to be 'protective', an office 'practical', a ball-room 'festive' and a church 'solemn'. When we visit a foreign city, we are usually struck by its particular character, which becomes an important part of the experience. (...) In general we have to emphasize that all places have character, and that character is the basic mode in which the world is 'give'. To some extent the character of a place is a function of time; (...) The character is determined by the material and formal constitution of the place".

[130] Peter Jordan describes in the volume Trends in Exonym use the usage and the significance of endonyms and exonyms in the construction of local identities.

Thus, in the process of naming places, three factors are involved, according to the analysis of Professor Peter Jordan: geographical space, social groups and names to be attributed (Jordan, 2011, pp. 11-12). Among these factors, the most important one is the social group. Within it, the names of places play an important role, being territorial markers. For a social group, the names attributed to the places to which that group (it may be the case of several social groups as regards the multiethnic regions or localities) is connected by culture, daily life or habits are called endonyms (Jordan, 2011, p. 11)[131]. Compared to them, exonyms do not express the membership in a certain social group, being rather an adaptation in the language of another social group of the name given by the native group and belonging to a culture specific to a certain area.

The names attributed to geographical places are engines of social relations and are the very core of the creation of the group identity and of the local identity. Here the argument may continue in the sense in which it is the name given to a place by a social group that is the instrument by which the social group defines its identity. It is a cyclic, dualist construction, in which the social group located in the space determines the very transformation of that space by its locality and its transformation into place. Thus, the clear distinction is made between oneself and the other, between space and place, between other already located spaces, namely between other two or several places.

Figure 35. An example of use of trilingual plates. We have, as in the previous case, the name of the locality in Romanian, Hungarian and German. Copyright: Alexandru Calcatinge.

Here is how, in a phenomenological approach of landscape, we must not avoid discussing about the importance of language and of geographical names in the creation of conscience[132] and of the membership in it. The awareness of the membership in a space, either geographical, architectural, existential or percep-

---

[131] For a better understanding, we recommend the study of the diagram suggested by Professor Peter Jordan in his study.

[132] As regards the conscience and its relation with the space or the geographical region, we provide further details in the chapter dedicated to this topic at the end of our thesis.

tual, is produced first of all when that space is given a name[133]. Naming a space either urban/rural or non-anthropised natural turns it automatically into a place in the human conscience and perception. At the same time, the space becomes a place once it accomplishes other functions besides the primary one of having a name, of being personalized and acquired (Tilley, 1994, pp. 18-19)[134]. In other words, we mention the fact that a space becomes a place once it has received a name, but it also has to comply with the characteristic of being located, as we will show hereinafter.

In the opinion of Botolv Helleland, Professor at the University of Oslo, there is a tight connection between the place name, landscape and identity. Thus, the argumentation is built based on definitions he gives to these terms, in order to be able to determine the meaning and the function of the place name. Consequently, the author defines the place name as a linguistic expression which describes the perception towards a spatial entity (Helleland, 2009, p. 26)[135]. This name has an important in defining the identity of a landscape (Helleland, 2009, p. 26)[136] in the conscience of the individual. Thus, a certain landscape possesses identity once it has been located, the reference to the latter being made by its name.

The name given to a landscape is only the first phase of the locality process and, as Tilley reminds, the practice of native Apaches in the North of America was to name places according to their description and physical characteristics (Tilley, 1994, p. 19). Thus, it appeared the locality of a certain space by its description and that of surroundings in its name. As Botolv Helleland himself states, the process of name attribution to a place is essentially a labeling of what may be perceived and observed by the individual in the environment (Helleland, 2009, pp. 26-27)[137].

---

[133] Here we can add the fact that in the case of the virtual space, too, as we will see in another chapter of this paper, the same rules apply. In case of Internet, we associate each address to a name. These addresses have the name of site, namely the place. It is a virtual place in the virtual space.

[134] Original quotation: "By the process of naming places and things they become captured in social discourses and act as mnemonics for the historical actions of individuals and groups. Without a name culturally significant sites would not exist, but only as a raw void, a natural environment. In a fundamental way names create landscapes. An unnamed place on a map is quite literally a blank space".

[135] Original quotation: "a place name is a oral or written linguistic expression that two or more people use with reference to a particular spatial perception".

[136] Original quotation: "Landscape identity may thus be seen as oneness between self and landscape as it is experienced by human beings in given circumstances. (…) I will also argue that a person's landscape identity (or place identity) is strengthened if he or she knows the place names of the area".

[137] Original quotation: "As a part of language, place names are first and foremost labels attached to one's surroundings, as useful and indispensable today as they no doubt were in prehistoric time. (…) They make part of one's identity both as linguistic terms and as identifiers of one's environments. Place names may be used to bring forth a wide range of mental and emotional associations – associations of time and space, of history and events, of persons and social activities, of oneself and stages in one's life".

Professor Peter Jordan presents three relations of the geographical space and of a group culture, in order to clarify the relation that place names have with its identity and with the cultural landscape. Thus, in the author's opinion, a group culture is based on natural resources in order to develop, it creates a space forming what we call cultural landscape and it equally builds the identity by what the cultural landscape offers (Jordan, 2009, p. 33). In the argumentation of Professor Peter Jordan, the important role of place names is obvious in the last two relations of the three presented relations. The cultural landscape is thus shaped by means of names, both at the visible level (by visible elements in the landscape) and at the mental level (since at the cognitive level an important man-space-place connection is shaped by labeling the located spaces) (Jordan, 2009, p. 36).

9.3 The concepts of chronotope and chronotopy in the study of the cultural landscape

The term chronotope was first introduced by the Russian literary critic and philoso-pher Mikhail Bakhtin. He used the term in literary studies. As we are already used to, a series if important terms used in philosophy and cultural studies are named by Greek terms. This is the case of the present concept, which designates the relation between time and space. The Russian philosopher imported it from the theory of relativity of Albert Einstein using it thus in a meaning suitable for his literary studies. The essay in which the concept was introduces by Bakhtin (1981) was written in the third decade of the past century (Brandao, 2006, p.133) and represented one of the important works of the Russian philosopher and literary critic in a difficult historical period. The novelty of the concept represented the thesis according to which space and time are inseparable in literature as well as in the daily reality.

9.3.1 Defining chronotope as a unit and chronotopy as a process

What is a chronotope? Mikhail Bakhtin defined it as "the intrinsic connectedness of temporal and spatial relationships that arc artistically expressed in literature" (Bakhtin, 1981, p. 84). At the same time he affirms that an important aspect for the literature studies in the study of this concept is the connection between space and time and the inseparability of the two categories (Bakhtin, 1981, p. 84), and „ these two categories constitute a fundamental unity, as in the human perception of everyday reality" (Bemong and Borghart, 2010, p.3).

Bakhtin did not give the concept a very clear definition, which determined its use in different branches. The chronotope received thus a series of meanings and different approaches, all these being in strong relation with the original description of Mikhail Bakhtin. Completing the above mentioned definition the author affirms: "in the literary artistic chronotope, spatial and temporal indicators are fused into one carefully thought-out, concrete whole. Time, as it were, thickens, takes on flesh, and becomes artistically visible; likewise, space becomes charged and responsive to the movements of time, plot and history. The intersection of axes and fusion of indicators characterize the artistic chronotope" (Bakhtin, 1981, p. 84).

We will continue our theoretical exposure showing the fact that a chronotope represents the element that enables the synchronic organisation of the connection points between time and space which structure the relations which define human activity. Being a temporal and spatial concept, chronotope is a dialocial concept and represents a specific social dualism. We thus extend the statement above with the thesis according to which chronotope is composed of a significate and a significant which determine the specific dialogism, which preoccupies us in the studies is cultural landscape. The significate is the very concept, the basic idea which determines the meaning, and the significant is represented by the material form, the image and its substance. The topic side of the used concept, its spatiality is determined not by the abstract, geometrical or universal meaning of the term of space but by the social side as a result of human action. Thus, at the level of the cultural landscape, the time-space system symbolises the (power) relation which mark and define society. The direct implications which can be noticed in the study of cultural landscape reduce to knowing the categories and types of existent chronotopes and which are essential to the practice of our profession.

For the study of cultural landscape the importance of the concept chronotope must be acknowledged through several approach directions. Before that, we have to take into consideration the fact that in the evolution of the concept cultural landscape the problematic of space appeared in a period in which the concept of time dominated the sphere of theories in philosophy, geography or cultural studies. Appealing thus to the contested characteristic rurality of the concept first introduced by Carl Ortwin Sauer of cultural landscape, we can easily deduce the intrinsic relation between space and time. In the way in which the concept chronotope is used in literary studies, filmography, documentaries or plastic arts, this can be used also in the studies of cultural landscape, being a permissive one and owns a certain conceptual flexibility in use.

Thus, a first approach involves knowing the fact that in the conception of Bakhtin, space and time are the social and historical components which unify the concept of chronotope, making it a dialogical one. Thus the art of dialogue in the exposure of ideas and feelings from Bakhtin's studies becomes important for us n the studies of cultural landscape. We discussed several times along our research about the dualism of existence and of the concept of cultural landscape as well as about the relations which build it. We can thus state that the dialogue nature vs. culture as well as the duality of existence and the complexity of the reading process (lecture) and the perception, lead to the meaning of the concept and of the process of the cultural landscape. The landscape has a cultural meaning which includes at the same time a temporal meaning, and the real and fictional time in relation to space are closely connected to what we call identity and specific. We thus tend to argue the thesis according to which the cultural landscape is a chronotope f meaning and sense.

Even if a century passed from the first concepts used in this chapter, cultural landscape and chronotope, it is impossible not to observe their validity and actuality in actual theoretical approaches. Appearing in close times but in totally different social, political, cultural, economical and ideological contexts, the two concepts permit a complex approach through their flexibility and apparent ambiguity. As we stated before, this very ambiguity determines the high interest of the study and make the mentioned concepts own a special contextual wealth. In this way Craig Brandist states that "Bakhtin, it seems, offers something to everyone "(Brandist, 2002, p. 1). The same author explains the influence that Bakhtin would have had adopting this denomination (the one of chronotope) (Brandist, 2002, p.123), and the explanation he gives to chronotope would be that it is the way to describe „ spatial-temporal relations in culture, and this acts for Bakhtin as the main link between aesthetic form and history" (Brandist, 2002, p. 123). At the same a valuable pointing out of Craig Brandist expresses the fact that every epoch presents different visions over the categories of space and time making thus a parallel between the way in which Immanuel Kant considered that the categories of space and time being apriori and transcendental while for Bakhtin these are only part of a objective and valid reality being thus partially transcendental (Brandist, 2002, p. 123). By the concept of chronotope the author present thus „ a historical typology of the architectonic forms of literature: specific works are expressions of certain historically specific senses of space and time" (Brandist, 2002, p. 123). The senses given to the categories and concepts with which Bakhtin operates, although being an adept of the Soviet formalism, can be understood knowing the theoretical influences which he had during his life. This is described by the same Craig Brandist, who presented the main philosophical influences which determined Bakhtin's works. Among this he underlines the importance Hegel had in lining out his philosophical profile, but at the same time he describes Neo-Kantianimsus, Lebensphilisophie, phenomenology, the theory of Gestalt of Brentano as well as religion as being very influential (Brandist, 2002, pp.15-24).

A second approach involves knowing two operating levels of the concept chronotope. Thus, according to a study made by the Vienna University regarding Bakhtin's chronotope two levels are being distinguished, which this concept operates on (Ganser, et. al., 2006, pp.1-17). The first level represents the way a text represents hispory (Ganser, et. al., 2006, p.2) and the second is the way in which the image of time described in a text and of space is represented, image which has to be built by any representation of history (Ganser, et. al., 2006, p.2). Time and space have not always had similar positions in the different modern or post-modern theories. It is known that in Modernity time preceded space but Post-Modernity debated about the latter one. The two became paradigms of the 20th century and the 21st century and in the studies of cultural landscape the historical and fictional time as well as place and space become paradigms which were always subjected to changes and adaptations

Thus, the use of new concepts in the process of understanding the cultural process is imperious and we consider chronotope as being one of them. The way we use a concept from the literary studies in the studies of cultural landscape offers us the possibility to emphasize the conceptual potential which the studied term has.

We have earlier showed what chronotope is, but in our presentation we will tackle also the term chronotopy as a process. In understanding the meaning of this term we can refer to the term cultural landscape is both a process and a perspective. Thus chronotopy is a process by the state of evolution and (trans)formation which it gives to the chronotope.

In order to facilitate understanding the chronotopy of the cultural landscape we will appeal to one of the works of Mireya Folch-Serra, Canadian geographer. Thus the chronotopic analysis in Mikhail Bakhtin's works, the study of the dynamic relation between the self and the other as well the methods of cultural analysis of the known Russian philosopher represented the main directions which Mireya Folch-Serra followed in his research on landscape as she considered that „ These are the foundations (the subtext) upon which the concept of landscape is laid out" (Folch-Serra, 1990, p.257). At the same time she states at the beginning of her work that „Bakhtin knew that human agents[138] wield

[138] The definition of the term human agency : Agency is a concept used in philosophy and sociology to refer to the capacity of an agent to act in a world. In philosophy, the agency is considered as belonging to that agent even if that agent represents a fictitious character, or some other non-existent entity. The capacity to act does not at first imply a specific moral dimension to the ability to make the choice to act, therefore moral agency is a distinct concept. Or: Human agency is the capacity for human beings to make choices and to impose those choices on the world. It is normally contrasted to natural forces, which are causes involving only unthinking deterministic processes. In this respect, agency is subtly distinct from the concept of free will, the philosophical doctrine that our choices are not the product of causal chains, but are significantly free or undetermined. Human agency entails the uncontroversial, weaker claim that humans do in fact make decisions and enact them on the world. How humans come to make decisions, by free choice or other processes, is another big issue. The capacity of a human to act as an agent is personal to that human, though considerations of the outcomes flowing from particular acts of human agency for us and others can then be thought to invest a moral component into a given situation wherein an agent has acted, and thus to involve moral agency. If a situation is the consequence of human decision making, persons may be under a duty to apply value judgments to the consequences of their decisions, and held to be responsible for those decisions. Human agency entitles the observer to ask should this have occurred? in a way that would be nonsensical in circumstances lacking human decisions-makers, for example, the impact of comet Shoemaker-

agency through language. We purport that this agency creates landscapes through metaphors and comparisons whose outcome is the building of roads, towns, and cultures. Landscapes, in this light, can be regarded as repositories of polyphony and heteroglossia: places where social, historical, and geographical conditions allow different voices to express themselves differently than they would under any other conditions" (Folch-Serra, 1990, pp.255-256).

In the above mentioned quotation we observe that the term heteroglossia appears being introduced by the same Mikhail Bakhtin in the year 1934. It refers to the way in which through language we can describe distinct units using a certain linguistic code. This concept can be associated to the term of dialect emphasizing the very structural complexity of language, which determines different meanings to the same object. Mireya Folch-Serra associates this concept to the mentioned landscape mentioning Yi-Fu Tuan, who considered the geographer a story-teller, a knower of the things, he talks about, but who during the story he becomes an active observer outside his story but directly implicated in its delination (Folch-Serra, 1990, p.257). We can extrapolate this description in regard of an architect and an urbanist through the very way in which their implication in the active processes of contemporary life determines the threat of the current story of both the present and the future. Thus „ Landscape becomes not only 'graphically visible' in space but also 'narratively visible' in time, in a field of discourses all attempting to account for human experience" (Folch-Serra, 1990, p.258).

What the chronotopy of landscape regards, Mireya Folch-Serra shows the advantage which chronotope has as an operative concept. This advantage is given by the equal importance which both time and space have in their structure (Folch-Serra, 1990, p.261). Event thus, the issue is which is the limit (temporal and spatial in the sense of threshold used by Josep Muntanola and presented later in the paper) from which the analyze of events / relations / processes starts, which determine chronotopy. This represents in essence both „the intersection of spatial and temporal sequences" (Folch-Serra, 1990, p.261) and „matrixes in which temporal events and spatial elements are equally valid and correspond one to another" (Folch-Serra, 1990, p.262).

The scientific discourse in the domain of architecture, urbanism or geography is different from the narrative discourse of the theories of the Russian philosopher Mikhail Bakhtin. Because on this fact Mireya Folch-Serra extrapolates certain types of chronotops met in Bakhtin's works in order to be used in the study of geography. We will refer to these as their structure at least the conceptual one is essential for our study of cultural landscape even more as it somehow shapes many of our past approaches. Chronotope, in the meaning given by

---

Levy on Jupiter. The definition is found on the site http://encyclopedia.thefreedictionary.com/human+agency and refers to the work of: Bandura, A. (2001). Social cognitive theory: An agentic perspective. Annual Review of Psychology, 52, 1-26. The Agency (1956). Describes the form of agency.

Mireya Folch-Serra, means the existence of certain situations both in the narrative development, as in the text, and in real situations as the spatial existence. Thus it is presented to us as an example of situation with territorial implications the human movement determined by migration, dwelling or spatial distribution. In these cases the spatial - temporal connection is evident as Mireya Folch-Serra affirms through „the reversibility of moments in a temporal sequence, and by their interchangeability in space" (Folch-Serra, 1990, p.262). The argumentation continues by the fact that „time becomes a quasi-perpetual time" (Folch-Serra, 1990, p.262) in the never-ending movement of the individual. It is mentioned a chronotope of the viewer in which the cultural roots of the one that performs the research are essential for understanding the researched object (Folch-Serra, 1990, p.262). The reference to "folkloric time" which appears often in Bakhtin's works is strongly legitimate commenting that along time due to the appeared changes on all the levels of human life „ the internal disintegration (bifurcation) of each element of the matrix becomes more and more intense" (Folch-Serra, 1990, p.263). Thus according to the Canadian philosophy, „ from the eighteenth century onwards a new feeling for time begins to awaken: the problem of time is elevated to the level of philosophy" (Folch-Serra, 1990, p.263).

Having representational qualities chronotope is the one that shows time and space with the help of narration within a written work. Even if it seems quite abstract around this concept a series of other concepts, ideas and ideologies revolve, which help for the best understanding regarding the analysis of some elements less abstract. In essence „the chronotope, functioning as the primary means for materializing time in space, emerges as a centre for concretizing representation, as a force giving body to an entire narrative whether a novel, an ethnography, or the analysis of a region. (...) Such is the representational significance of the chronotope. By investigating landscapes through the optic of chronotopes, we always arrive, in the final analysis – even if passing through a lengthy series of mediating links – at the human voice, which is to say we come up against the human being" (Folch-Serra, 1990, p.263).

Thus the narrative discourse takes place at the level of cultural landscape, too reaching its maximum at the level of the contemporary city. We can easily deduce the fact that his specific narration and dialogism have always been suppressed to ideological changes and to the paradigm. And we consider that the narration of the city and the territory, their textuality is best pointed out through the space of temporality or as we will show in this chapter through chronotopy. This as "to pass through time and space is our very way to be as human beings" and "the question regarding passing through time is put (among others) by philosophy, the one regarding passing through space falls upon (among others) architecture and geography" (Ioan and Mihali, 2009, p.162). This is why the study of such a concept as cultural landscape brings us to the confluence (at least) of the three subjects.

9.3.2 The cultural landscape from chronotope to the social dialogism

The social space, non-abstract in its construction, has represented since the end of the 19th century and the beginning of the 20th century the very playground, a real laboratory of abstract expressions and visionary projects as expressions of the new cultural, social and economical ideologies. All these visions were captive as at the same time part of an extremely rapid evolution, which characterized the last century. Everything happened as Kai Voeckler used to say, too at the speed "turbo". Referring to this whole period, especially to the after-war one, we can refer to a "turbo-culture", "turbo-architecture" and "turbo-urbanism" (Voeckler, 2008, pp.8-11). And in this constant rush characterized by movement and speed, the need to slow down and have a break (but not contemplate) in order to understand what we call cultural landscape is absolutely necessary. Break which is necessary in order to write as Bogdan Ghiu affirmed in an interview taken by Ioana Crăciunescu within the project of "Writers on the Royal Path" presented also by TVR Cultural in the year 2008.

In this whole "turbo" speed William Mitchell affirmed that „ it was a short step from reading nature – which is utterly indifferent to human needs and purposes – to reading artifacts. And artifacts do have intentions behind them" (Mitchell, 2008, p.vii). These intentions are those which determine the messages which artefacts have as well as the power relations and the dialectics we mentioned above. These artefacts are chronotopes, and this can be deduced along our argumentation. And each of them has a significate and one significant, that is an idea as well as a material form. Still the name of artefact, as Mitchell uses it, is not quite eloquent. As we could observe in the previous chapters, the term artefact was assimilated to the process of understanding a cultural landscape. The extended understanding over this term comes from Mitchell, too, who emphasises its dualism by the fact that artefact has a double role both functional and symbolical (Mitchell, 2008, p.x). By these the architect extends the meaning of artefact to the elements of a construction as well as to the whole construction. And artefacts, as we discover through Mitchell's thesis are found as characteristic elements in the study of cultural landscape. The argumentation which William Mitchell builds regarding the role of artefacts (even the construction ones) is useful for the study we perform, in the context in which the need to understand the cultural landscape, to read it and correctly frame it in the creative process of our profession is greater and greater. Thus Mitchell affirms the fact that „it helps to have some way of separating physical and symbolic tasks. In other words, we need systems of abstract, dematerialized, cost-free artifacts that can serve, in efficiently specialized ways, almost entirely as carriers of messages" (Mitchell, 2008, p.xi).

Being today in the sphere of theory in architecture and urbanism dominated by West-European continental philosophies, what the Russian philosopher offered can represent an absolutely necessary novelty. According to Michael Holquist (2002), the philosophical roots of dialogism are found ion the need to fill the emptiness between material and spirit appeared at the adepts of Neo-Kantianism (Holquist, 2002, p. 16)[139]. He describes then the way in which Bakhtin's dialogism differs as theory of knowledge from the other theories of this kind. Holquist affirms the fact that the foundation on which dialogism formed is represented by the non-identity of mind and the world (Holquist, 2002, p.16), and for Bakhtin this is the very ground of the levels of non-interactive identity (Holquist, 2002, p.16). Thus Bakhtin's way of thinking is considered a meditation on knowledge, based on dialogue, due to its non-unity. Talking about unity, Holquist's discussion about the theory of Bakhtin's knowledge reaches the shere of knowledge and its unity. Thus in dialogism „consciousness is otherness" (Holquist, 2002, p.17). Still it is subscribed to a relation of differentiation between a centre and the rest, what is different, what is something else, besides of that centre. And for Bakhtin this centre is, as Michael Halquist descries, a relative term with no claim of generalization (Holquist, 2002, p.17).

Continuing the analysis of the above mentioned, we will refer to Carl Ortwin Sauer's binomial according to which nature + culture = cultural landscape. Its simplity can be a trap in the conceptual study. This binomial transposes in a certain dialogism, in which terms as difference or centre and conscience become important variables of the ecuation. This gets thus new meaning. Culture involves conscience and determines specific and identity. The dialogue culture - nature is thus a dialogue of the social and history, a dialogue of the individual and the world, a dialogue between me and the other, between historic and fictional time, a dialogue of spaces and places. The difference is observed from the first contact with Sauer's equation. The very contraposition of the two terms of culture and nature shows a clear difference, and the centrality of one towards the other became obvious once with the evolution of the concept oscillating. Culture became the centre which, as we mentioned above, is not a absolute one. For Bakhtin culture gathers three important domains as science, art and life (Tihanov, 2000, p.53). At the same time, as Galin Tihanov stated, Bakhtin's inclusion of life as part of culture in his theory shows that even if the two are different, culture is based on life (Tihanov, 2000, p.54). We have to state the fact that Bakhtin was sceptical about considering art a privileged field

[139] The origianl quotation: "After the death of Hegel, this gap became increasingly apparent in the growing hostility between science and philosophy. Dialogism, then, is part of a major tendency in European thought to reconceptualize epistemology the better to accord with the new versions of mind and the revolutionary models of the world that began to emerge in the natural sciences in the nineteenth century. It is an attempt to frame a theory of knowledge for an age when relativity dominates physics and cosmology and thus when non-coincidence of one kind or another—of sign to its referent, of the subject to itself- raises troubling new questions about the very existence of mind".

of culture (Tihanov, 2000, p.54), the latter one was created „ in a process of constant redefinition of forms and an exchange with life in which existing forms are appropriated and new ones are created" (Tihanov, 2000, p.56). Mikhail Bakhtin speaks about these forms from the perspective of th aesthetic theory, but for our study, his theory about the act / action is important for determining the meaning, perception, reading and interpretation.

For the Russian philosopher all important activities (naming here the theoretical way of thinking, the exposure and the historic description as well as the aesthetic intuition) have in common the fundamental separation between contents and the sense of an act / action and its unique existence and experience (Bakhtin, 1993, p.2), what leads us to the confrontation, according to the author, of two impenetrable worlds, the one of culture and the one of life (Bakhtin, 1993, p.2). The latter one being the only world in which we live and in which the culture world is included by the processes of knowledge, creation and thinking being the only world in which „ the acts of our activity are objectified and the world in which these acts actually proceed and are actually accomplished once and only once" (Bakhtin, 1993, p.2). The author continues, „An act of our activity, of our actual experiencing, is like a two-faced Janus. It looks in two opposite directions: it looks at the objective unity of a domain of culture and at the never-repeatable uniqueness of actually lived and experienced life" (Bakhtin, 1993, p.2). We can thus state that the existence involves a continuous dialogue, as well as experience.

Regarding science and technology Bakhtin introduces the problem of necessity[140]. Thus „the ought is capable of grounding the actual presence of a given judgment in my consciousness under given conditions, i.e., the historical concreteness of an individual act, but not the theoretical veridicality-in-itself of the judgement" (Bakhtin, 1993, p. 4). For a better understanding the author continues stating that no special theoretical necessity is needed, as the trustworthiness or stating the truth is the very necessity of thinking (Bakhtin, 1993, p.4), and necessity appears only in relation with the act of thinking and the truth itself, and this correlation represents a unique historic moment (Bakhtin, 1993, p.4). Thus technology is suppressed to a own law under which it unfolds, it being deviated form its initial necessity, becoming in the vision of the Russian philosopher a threat and a way which can serve any time a destructive force (Bakhtin, 1993, p.4). This vision on technology and its malefic finality was influenced by the suspicious and difficult environment in which Bakhtin lived (McCarthy and Wright, 2004, p.56) and the censorship to which he was constantly subjected by the system.

I stated earlier that dialogue is characteristic to existence and experience. The latter one is an important subject in the study of cultural landscape and mainly in the study of human and cultural geography, but also in many studies of ar-

[140] Or the term the ought in English or Sollen in German as it appears with Kant - the necessity of an action seen either as a action medium or as a purpose being equal to wish so must = will.

chitects, sociologies and urbanists. Even more as the need to understand certain tings is so great through the fact that the result we offer by the proposed constructions and solutions influence the experience of others. According to this both science, art and life as components of culture, as Bakhtin stated, influence and are active variables for the existence.

I broadly presented a series of ideas and theories regarding experience in our last approaches, and now it is time to bring some additions regarding the subject debated in the present chapter and under this occasion to extend the study to new theories which can better clarify what the cultural landscape represents, both as an operative concept as well as a reality exposed to daily experiences, subjected to interrogations, contemplations, strong analysis and theorizing.

For a better understanding, we will further extend the exposure of what we consider is the dialogism of cultural landscape. Dialogue implies a simultaneousness of actions and the existence of a asymmetric spatial - temporal duality, which produces relations between the implicated elements. The dialogue itself represents a relation. In order to understand this, Michael Holquist's exposure presents the way in which simultaneousness has a fundamental role in Bakhtin's dialogism (Holquist, 2002, pp.17-19). Thus, the author states that dialogue helps to understand fundamental relations, which for Bakhtin were considered artificial and for which knowing the self and the other through dialogue was primordial (Holquist, 2002, p.18). Thus, „ self/other is a relation of simultaneity. No matter how conceived, simultaneity deals with ratios of same and different in space and time" (Holquist, 2002, p.18). At the same time the author states that dialogue does not tolerate suppression (Holquist, 2002, p.18), referring thus to Hegel's dialectic. In this meaning, considering the term of suppression, which comes from the German term aufheben or the English one sublation, and knowing the fact that in the philosophical studies this means sequence of ideas or concepts which complete the previous ideas without eliminating their principles and meanings, contributing to the fulfilment of a complete understanding, we can deduce that fact that dialogue is part of a pragmatic discourse of the meaning/sense. Even if dialogue does not tolerate suppression, this phenomenon comes later on as a result of dialogism among the essential subsystems as in the case of cultural landscape, the social, historic or mental side. Suppression implies enrichment and, at the same time, a kind of negation together with conservation. It is a term quite difficult to understand, but in Hegel's dialectic he kept the whole sequence of the above mentioned meanings. Thus, through dialogue the initial meaning is never lost it is always enriched, even if the meaning is changed. The richness of the term is given by the name in German, where it means both negation and conservation, in which the meaning (of a term, concept, a reality, etc.) is saved and kept once with the gradual enrichment (distinct temporality of the chronotopic construction) of the meaning. The simultaneity of the use and existence of this term, which has two opposable meanings, is the very essence of understanding the simultaneity of dialogue and its complexity (Palm, 2009, pp.9-13).

The dialogue of cultural landscape is part of a discourse of dualisms as culture / nature or space/time, and space/place, as we observed earlier. In this way the extending of the meaning for the notion of time used along our exposure needs supplementary explanations. Following the analysis up to now by using the term chronotope in our studies about cultural landscape we accept the spatial / time framing of the concept. This is history, present and future at the same time. The notion of time is maybe one of the most difficult notions in philosophy, and the purpose of this exposure is not to bring the analysis of the term in this direction, of the philosophical studies, but to facilitate understanding the concept of chronotope, used in literature, which we attribute to cultural landscape, which is a concept of cultural geography.

We know that time is subjected to space in contemporary studies. What was called from the very beginning cultural landscape was equally subjected, at least ideologically, to time and space. Both culture and nature are each a spatial-temporal dualism. The topic change has always implied a temporality. Being a witness of passing of time and spatial adaptations and changes, cultural landscape is in our acceptance an ideal chronotope. In sustaining this, we refer to the vision of the city and the landscape as a text or a book[141], and reading these implies experimentation, perception, understanding spaces, and all these experiences imply the existence of a temporality. All these imply a past, historic, present time, the experience lived at the moment of perception as well as a fictional, future, imagined, visionary time. All these states are connected not only to a certain space, which is experimented, felt, lived, thought, but also to a past, present or future time.

Let us remember the term palimpsest that we used when defining the concept of cultural landscape in our past exposures. It detects both the stages of topic evolution and those of chronological evolution. The palimpsest is a spatial-temporal dualism in the case of cultural landscape. Passing of time, which is the chronological variable of this duality, joins the topic sequence of different human interventions within the natural frame. In essence, we consider that this fact meant for Carl Ortwin Sauer the concept of cultural landscape by the equation proposed in the first half of the past century. This is how all the theoretical exposures up to now managed to converge towards a finality, which can be only useless up to now. The used terms and notions such as the given definitions are used in framing the ideal chronotope represented by the cultural landscape. It is easy to observe how the concept used too many times in a simple way of-fers a disarming complexity.

Coming back to what the term time, we appeal to what Oswald Spengler said, who, comparing the two paradigms, space and time, concludes that time is organic an attribute that space does not have (Spengler, 1944, p.122), it being simply a conception, an idea, a notion (Spengler, 1944, p.122). At the same

141 I analysed this aspect in the book Visions of the Real. An architect's approach in cultural landscape studies published in 2011 in Viena.

time he considers time as „ a word to indicate something inconceivable, a sound-symbol, and to use it as a notion, scientifically, is utterly to misconceive its nature" (Spengler, 1944, p.122). Bearing the same idea in mind, for Spengler man is directly aware only of the space he is in, not the time, this being „a discovery, which is only made by thinking" (Spengler, 1994, p.122). This is how, Spengler concludes, by naming time, we managed as people to dominate it (Spengler, 1944, p.123) and still we did not manage to fully understand it. Spengler as Heidegger (2006, p.371) refers to Augustin's works referring to time and its understanding. Thus, Heidegger strengthens the certainty that time is a term, which is complicated and difficult to explain even by the most intelligent people (Heidegger, 2006, p.371), which makes „us not to know where to look and where we have to search and find something like time" (Heidegger, 2006, p.371). Time present a sequence, as Heidegger demonstrates in his analysis of Aristotle's conception of time (Heidegger, 2006, pp.373-408), and referring to this Spengler affirmed that we ourselves as human beings are time (Spengler, 1944, p.122)[142].

At the same time we continue our exposure mentioning the fictional time from Bakhtin's theories and the fictive experience of time introduced by Ricoeur (1985, pp.100-153). Even if both notions refer to the literary study, the connection with the studies of cultural landscape can easily be made. Paul Ricoeur affirms that o [literary] work can be at the same time both closed and open and „the fictive experience of tie is this temporal aspect of virtual experience of existing-in-the-world which the text proposes" (Ricoeur, 1985, p.100), such as, according to Heidegger, time implies a sequence of the former and the later (Heidegger, 2006, p.411), and this determines its infinity (Heidegger, 2006, p.418). The fictional time is the imagined, thought time, it is maybe the time of flaneur, or better the tie of creation, thinking, planning (with which we are accustomed as architects and urbanists). It is different from that "time of world's order" (Heidegger, 2006, p.420), it is a time of the text in literary studies (Ricoeur, 1985, p.100) and of creation and experience in the studies of cultural landscape. It is a dynamic time of human subjectivity and of interpretation and representation.

Thus, man's intervention by what we call construction [of the cultural landscape] is a process of social dynamic by identity, specific, character and culture, and of history, by the dialogical relation between the real world of existence and the fictional one of subjectivity, experience and singular human perception. The social contract of the individual implies a symbiosis man-nature which is characteristic to any culture. The need of a decent life and living together with other individuals as well as the need to adapt to the conditions of the environment and the whirligigs of nature lead to the orientation of human

[142] The original quotation: "We create it as an idea or notion and do not begin till much later to suspect that we ourselves are Time, inasmuch as we live". The note is presented as follows: "The sensuous life and the intellectual life too are Time, it is only sensuous experience and intellectual experience, the "world," that is spatial nature to Time, see (Vol. II, pp 403 et seq)".

being towards a better adaptation to nature. Thus, the harsh and unforgiving reality lead to the adaptation of human intellect in order to construct and offer shelter and safety. Creativity, the capacity to think and innovate determined the superiority of the human being. The dialogue man - nature thus implies understanding by the individual of a specific language in order to be able to "communicate" with nature. In this context identity and the local specific are born. The cultural landscape does not exist only in a natural context (being also culturally constructed), which is the scene of the whole action and at the same time the main actor, as it does not exist without the so-called culture. Let us remember what Roger Scruton said "as culture and the quality of being a member of society are one and the same thing" (Scruton, 2011, p.9). Thus, in his work he teats the way in which culture represents searching for identity, what had been earthed since the first definition of culture given by Johann Gottfried Herder in the 18th century (Scruton, 2011, pp.9-11).

We cannot talk about the individual without talking about the environment in which he lives, which determines through its components its existence and experience, its culture and thus the identity and local specific are born. This is how Roger Scruton affirms the fact that this search for identity that we mentioned above is a "way of the being-together, (...) a way to request space in a public world" (Scruton, 2011, p.11). The cultural landscape presents thus a problem of perception and representation of human social dynamic. These two characteristics represent variables of its specific dialogism. The model of the cultural landscape starts to be shaped as a chronotope of meaning and sense. Both processes, the perception and representation one, assume the use of imagination, feelings, memories as well as thinking and knowledge by the individual involved both in production and investigation of cultural landscape.

While building our argumentation we have to remind the fact that cultural landscape is a concept which is used by the inseparability of space and time as Mikhail Bakhtin used the concept in his literary studies. The spatial-temporal inseparability is closely connected to the culture-nature inseparability within the study of cultural landscape. In sustaining and demonstrating the observation that the theoretic discourse concentrated on certain circumstances on an overexposure of landscape as an important component in the perception and reading process of cultural landscape, we want to mention that what Hegel said about nature, that "nature is to be considered as a system of stairs, each of them coming from the other one, but not meaning that it is naturally created from the other one, but that it is born out of the inner Idea, which is the basis of nature. The movement of nature's Idea consists in going from its directness in its self, in outrunning itself and become Spirit" (Hegel, 2006, p.184). This way we can continue the exposure and remind the fact that "space and time are abstractions being-in-everything or pure form, the pure intuition of nature - Space, thought (Gedanke) of general diversity generally indifferent, Time, the thought (objective) of the negative unity or pure becoming" (Hegel, 2006, p.184).

Post-modernity brought forward the discussion about space which is correlated with the never-ending search for identity in a world of dualisms and multiculturality, in which tie seems to be forgotten. In his exposure about modern culture Roger Scruton talks about a common culture and a high culture (Scruton, 2011), and in his opinion the third conception about culture refers to a popular culture (Scruton, 2011, p.11) which together with globalisation lost its inherited side dominating only the acquired one (Scruton, 2011, p.12). In this way we can affirm that the first among the dimensions of time given by Hegel, the one of past, was forgotten. We can not talk about a total accidental amnesia of society. The frenzy everybody was overwhelmed with once with the apparition of modernism and with the radical paradigm changes in cultural and urban studies lead to redefining several fundamental concepts which grounded post-modernity. So we can understand why concepts from almost a century ago came to attention to the new post-modern theories. The return of time as important element in the study of cultural landscape has to be seen as a necessity, an antidote to the expansion of doctrinaire spatiality. Thus chronotope [time-space] is in our acceptance represented by the cultural landscape as being its ideal and total form. The supremacy of space will always exist, but we have to always keep in mind the inseparability of the two categories, as the concept chronotope indicates.

As architects, urbanists or geographers we operate with the topic category and use all its dimensions. Time is the connective category, always present, inseparable of the one of space, being considered its third dimension. As writers were creating in their artistic works spatial and temporal images of reality, space creators, total artists, as architects used to be called are those who build scenes of the reality lived by each of us. These are support scenes of life, emblematic frames for the life, the experience and the becoming of everyone. It is worth mentioning the fact that Mikhail Baklhtin showed the important character of representation which chronotope has (Bakhtin, 1981, p.250)[143] being according to the author the main means of materialisation of time into space in his literary work (Bakhtin, 1981, p.250)[144]. Ivar Tormod Berg Ørstavik (2005) emphasizes that „Without a time-space system of some kind, our conceptual perceptions and constructions would have no ground to stand on and no time to exist in. Thus, chronotope as an analytical category directs our attention towards a basic and intrinsic foundation for our imagination" (Østravik, 2005, p.425). The term of analytic category appears which Østravik proposes for the term of chronotope emphasizing the fact that it is used in oder to „ describe the imagined time-space relationships or systems that are constructed together with these

[143] Original quotation: "We cannot help but be strongly impressed by the representational importance of the chronotope. Time becomes, in effect, palpable and visible, the chronotope makes narrative events concrete, makes them take on flesh, causes blood to flow in their veins".

[144] Original quotation: "Thus the chronotope, functioning as the primary means for materializing time in space, emerges as a center for concretizing representation, as a force giving body to the entire novel".

fictional worlds in order to sustain them" (Ørstavik, 2005, p.425).

Far away from the fictional worlds of literature, the cultural landscape offers other worlds which we can consider fictional themselves. In our argumentation we emphasize once again the fact that the built environment is the result of creativity, imagination and knowledge. This is a product of community, or at another level of different member individuals of one or several communities. Thus, the belonging to a certain social group, the discovery of the identity we mentioned above, the common acquired and inherited culture, all these are the basis of a spatial-temporal evolution of what we have been calling cultural landscape since a century. Thus, according to Brian Elliott (2010, p.xii)[145], community and the place in its dialogical meaning by which we tackled the study of cultural landscape in this work are two inseparable components. Just as space and time. Reminding what Spengler (1944) said referring to the fact that we ourselves are time, this way the dualism community-space represents in extremis the dualism time-space.

The social dualism, besides what we have mentioned above, means also the duality of the existence nature-society, not just community-space. It is worth following the way in which professor Alexandru Sandu binds the term cultural landscape with modern dwelling (Sandu, 2007b, p. 27), or as, Peter Jackson (1989) underlines the fact that from the first interaction of Carl Ortwin Sauer's concept he drifted the attention from the social component of environment to the physical one (landscape) (Jackson, 1989, p.14). This brought strong critics on the concept of cultural landscape, but it represented the ideal basis for the evolution of the concept along the century. We consider this evolution a logical and necessary one, which brought the concept to the maturity it has now.

The antithesis culture - nature lost from its conceptual vigour it had in the past, or in other words from the power with which it was imposed through framing the first definition of the concept of cultural landscape. The discourse became a social one, as upholder of the modern and post-modern paradigms, which marked the last century. The dualism community-space and the one between nature and culture represent the very roots of social dualism we tackle. The latter one is characteristic, as we will later show, of the operative concept of cultural landscape as it is used in the contemporary professional literature. The social dualism is a dualism of moral values and materials expressed in a world in confusion and disorientation. In this way we can underline the fact that a critic point has been reached in which the dissemination of the above mentioned val-ues emphasize confusion determining moral, spiritual and material disorder which reign over the contemporary society. The dialogism of the cultural landscape transformed along time to critical dialogism that led to a decontextualisation of space in relation to time.

[145] Original quotation: "What binds the two parts is a basic assumption, namely that community and place are intimately connected such that the one cannot be adequately understood without the other".

From the point of view of social dualism that cultural landscape represents, this de-contextualisation is the heart, which, both at the urban and peripheric or territorial level, determines the change of paradigms which have been at the basis of evolution till now. In this full process of transformation, culture became an important element in the social equation, and together with the world economical and political changes has the role of a catalyzer.

The de-contextualisation of the cultural landscape implies its reinventing regarding the way to understand and use it, and dialogism as art of the dialogue and of expression of ideas is the essential property of cultural landscape in the process of its understanding and (re)inventing. The critic moment was reached when space became the universal paradigm of social, political and cultural discourse in the urban frame (but not only here). Thus, space contributed to a cultural contextualisation, being the mirror of social and cultural identity, as John Agnew affirmed, that spatial sensitivity represents the very basis of cultural differences (Bonnemaison, 2005, p.14)[146]. Time, as we previously affirmed, managed to exist as a critical element (here meaning essential / necessary / indispensable) of topic evolution, as the third dimension of space. Without the factor time in this duality, cultural landscape would have remained a metaphor. The de-contextualisation of this second element is produced through what history produces.

Thus according to those affirmed by Alan Baker (1996, p.1), time and history are those which produce the geographic personality, which was introduced by Paul Vidal de la Blache and which dominates the way in which human interaction with environment defines uniqueness, identity and specific and the fact that any description and understanding of the landscape have a historic ground (Baker, 1996, p.1)[147]. We mention here the fact that the term landscape is one imposed along time by cultural reasons and in itself by cultural processes. We can thus extrapolate the affirmation considering the whole landscape as a cultural construction. We can do this due to the fact that a mental construction of what every individual observes around himself[148]. It is thus a cultural product and process as well as a social construction. It's worth mentioning the fact that in the study of cultural landscape it is important what William Dodge called "traditional cultural properties" (Dodge, 2007, xi) among which he enumerated

[146] In the preface of the book of Joel Bonnemaison called Culture and Space. Conceiving a new cultural geography. Original quotation: „it is the spatial sensibilities of different groups that are at the root of their cultural differences".

[147] Original quotation: "Vidal de la Blache believed that geographical personality grows through time, that the interactions of people with their environment through time produce their individuality and distinctiveness of a landscape, locality, region of contry. It follows logically that geographical descriptions and understandings must be historically grounded".

[148] Here I find it worth underlining the fact that being at an international conference dedicated to cultural identity, the problem of contraposition of the two terms, nature and culture, was discussed. The argument from many important geographers was that the connection of two terms can be a pleonasm as i tis accepted in the interantional sceintific community (of geographers) that landscape is in essence cultural by the way it is perceived, understood and used.

the natural landscape together with constructions, traditions and other components which are today part of the cultural landscape (Dodge, 2007, p.12).

In conclusion we can affirm that cultural landscape, in framing its existence under the form of an ideal chronotope and a social dialogism, is the spatial representation (topos) of the sequence (cronos) of interactions and social relations determined based on the cultural processes.

We discuss again about the term relations regarding cultural landscape and the fact that „ The interactive links connecting people and landscapes, whether behavioral, social, or symbolic, permit identification of embedded landscape histories" (Dodge, 2007, p.12). At the same time it is argued that cultural landscape includes the unmodified landscape (Dodge, 2007, p.12) emphasizing the symbolical side of it.

Completing what we have so far argued, we will take into consideration the possibility to see the chronotope of the cultural landscape in at least two different ways. First by the way it determines the relation between the image of time and of space, and the second one by the experience [fictional] of time and space as a temporal aspect of spatial existence. Thus, at topic level, cultural landscape can be represented by both the results of human intervention and by natural landscapes with a strong symbolic character. Our research will be grounded mainly on the first way to see the chronotope of the cultural landscape, especially from the point of view of our profession, and the two other ways mentioned above are inseparable and interconnected, each human intervention having a symbolic or mental load for a community or a part of its members. At the same time the symbolic load is given first by the "author" of the intervention, as we will see in the next chapter of this thesis.

9.3.3 Chronotope and chronotopy within the creative process and human intervention

According to Raymond Brown and Peter Renshaw (2006, pp. 247-259) even in the educational processes the concept of chronotope of Mikhail Bakhtin can surely be applied as it „ provides a way of viewing student participation in the classroom as a dynamic process constituted through the interaction of past experience, ongoing involvement, and yet-to-be-accomplished goals" (Brown and Renshaw, 2006, p.247). In the same way, completing the above mentioned argumentation, Anna Louise Soutar said that „Without it being named as chronotopic, the evoking of human responses to axes of time and space is a focussing technique used in design studio where the students are encouraged by task and dialogue to recognise the impact of these sense-driven qualities on one's deeper level of imaginative understanding" (Soutar, 2003, p.71). In the studies of Raymond Brown and Peter Renshaw, we can observe the way in which the chronotopic analysis is applied in the case of general school pupils (Brown and Renshaw, 2006, p.250). The two consider the way in which children combine in the learning process both the real side of life, the one of the past and the present, and the imaginative one, of the future. Thus the chronotopic analysis is based on "the interaction between the past, the here and now and the possible and the imagined" (Brown and Renshaw, 2006, p.251). Extrapolating the above mentioned at the level of university studies (especially in the field of architecture and urbanism) we can observe the fact that in these higher schools the accent is laid on an objective knowledge of the norms, theories and laws, as well as on a intuitive one, which is "subsidiary to our conscious knowledge" (Soutar, 2003, p.74).

In her work Anna Soutar appeals to the works of the architect William Mitchell, which she considered the one who determined her to approach a study and a chronotopic thinking on architecture and the process of education in architecture. What the author mentions, which is in fact known by architects and urbanists, is the way in which "the electronic expression ways produced modified realities of time and space, so that the real time became a dimension; and the duration which shortened either fragmented or compressed into another one" (Soutar, 2003, p.79). Indeed this is one of the actual realities provoked by the technological evolution, and the education process in architecture and urbanism suffered some changes. The chronotopy of the creative process is obtained on informational way, because, as we can see form the above mentioned, it takes place an interaction of the past (memory and experience) with the present (the time of the creation) and the future (by the way to offer an exploitation of the creation through the virtual reality). Thus the time of creation is real, it can host at the same time an instance (that snapshot) of the future time ob construction, or as professor Augustin Ioan called "the tie after creation" (Ioan and Mihali, 2009, pp.87-97). The artistry of the original artistic expression which characterizes architects and urbanists transformed into the artistry of using the modern technology and transpose the narrativity specific to architecture as well as the intrinsic way of its spatial-temporal connection in a virtual world. The presentation of the finite product transforms the creative process into a chronotope by the way in which along the drawings specific of the profession we can create on the computer a virtual world which gives us a new way to "feel", perceive and read the cultural, urban and rural landscape. But this whole artistry in handling a tool must be supplied by the way the students (or the architecture and urbanism graduates) have to "find themselves within their own culture" (Doicescu, 1983, p.42).

Anna Louise Soutar (2003) proposes a study regarding the way in which technology creates chronotopes in the process of architectural creation. Thus, for a better understanding of the process, except of what we propose in our study and besides what we analyzed in the previous chapters regarding the virtual world and the study of cultural landscape, the work of the above mentioned author is worth being read. At the same time, completing what we analyzed in the previous chapters and guided by professor Alexandru Sandu, a deeper incursion of the way in which new technologies determine new way to experiment the everyday, we recommend also a deeper study of William Mitchell's works. We will mention him only for the chronotopical construction we build in this work.

One of the corellations William Mitchell makes with the spatial-temporal construction of the way of urban perception and with the narration type (reminding us of Mikhail Bakhtin) the city proposes is connected to the influence the new technologies will have on the way in which we will perceive the environment we live in (especially the urban one). The well-known American architect says that „these technologies create a new relationship between the city and its users. They create a new narrative of the city. And I tend to think that these narratives that unfold as you move through the city and the narratives that are mapped onto the city are very important for urban design. Narrative makes a city comprehensible" (Mitchell, 2007, p.408). He continues his exposure reminding that „ novels and plays basically provide experiences of situations and narratives and plots that in some sense are culturally engaging" (Mitchell, 2007, p.408).

We consider that the author's analogy with the literary work or the theatre is not by chance, especially knowing already the theory regarding the literary chronotope of Mikhail Bakhtin. As in a literary work, the narration is the one that gives birth to chronotopes, those structures which bind time and pace in a dual relation in which both categories acquire equal (narrative) importance. The same situation is within the creative process where the narration of environment, either urban or rural, (generally accepted as cultural landscape) is the one which gives birth to the chronotopes studied by us in this work. Completing those mentioned in the previous work, the dualism of cultural landscape at the urban level is given by the way in which "the virtual world conquers the physical one and the physical one conquers the virtual one leading to the development of some ambiguous areas" (Mitchell, 2007, p.409). As I affirmed with other occasions (Calcatinge, 2011, pp.105-110) this fact of fusing of the two worlds is translated by the trial of the contemporary individual to satisfy almost schizophrenic needs through transforming the real show produced by the urban experience into a virtual one, which facilitates other kinds of adventure and experience, another kind of knowledge and another kind of appeal to the memory.

But the major difference between the narrativity the creative process imposes based on technology as a too or the "classic" one is shown by Kai Hakkarainen (2010), who sarguments the fact that human mind is not accustomed with the learning methods based on technology being in full process of adapting to these tendencies (Hakkarainen et. al., 2010, pp.1-7). Regarding the application of chronotopy in this case, the authors affirm that the interpretation of the concept by learning through technology is resumed to the use of knowledge accumulated in the past for solving actual problems. Even if memory appeals to experiences lived with the help of classic tools such as paper, pencil or book, solving the actual tasks, even if being the same as the past ones, are solved by using digital tools, from planning software to reading electronic books (Hakkarainen et. al., 2010, p.1). Cognitions being interpretations between outer and inner fields, the classic tools are associated to modern outer representations, which are called by Merlin Donald external memory fields (Donald quoted in Hakkarainen et. al., 2010, p.1). These outer representations are considered to be "part of dynamic semiotic frame which intersects the mental, virtual and social lines of the activity area" (Hakkarainen et. al., 2010, p.1).

The chronotopy of human activity over the environment is easily understood if we come back to those presented at the beginning of the chapter. We thus come back to the term of chronotope with a series of additions in order to help along the present work. What I deliberately forgot to present in the introduction of my work will be proposed now in the construction of the way to see the cultural landscape as a result of human intervention. We will perform this differently from the already known approaches in this field. Still we will bear in mind the fact that we acknowledge the contributions brought by cultural geography, urban studies, sociology, ethnography, philosophy and cultural studies to creating the concept of cultural landscape, but from the perspective we approached in the present chapter we consider the need to shape the chronotope of the cultural landscape to be an important ring in the chain of specific theories.

Mikhail Bakhtin, in his well-known Forms of Time and Chronotope in the Novel that we mentioned at the beginning of our work, presents two categories of chronotope, the major and minor ones. But besides this classification, the author underlines the fact that there can be several categories of different complexities, difficult to be classified. So it is impossible for us not to be "impressed by the representational importance of the chronotope. Time becomes, in essence, palpable and visible, the chronotope determines narrative events, makes them get a form" (Bakhtin, 1981, p.250). The chronotope is thus the main "instrument" for materializing time through space (Bakhtin, 1981, p.250) and owns means of unique representation (within the literary work). As Tim Cresswell said regarding the work of James Duncan, landscape can be seen as a text (a literary work in order to make the connexion with the Mikhail Bakhtin's theories), or even more such as a 'material symbolic system in which people live" (Cresswell, 2008, p.6). He presents at the same time a kind of communication system, a space which by using a proper language leads to producing social and cultural relations which determine the built space. This is a chronotope in which space and time are prefigured.

Mika Aaltonen built the thesis according to which there is a chronotop(ic) space (Aaltonen, 2007, pp. 58-62) regarding the process of human creation. This process, that the author considers a strategic one is based on two important "intellectual" aspects. Thus the first assumes that „ the nature of a project, its work and the strategic landscape where the work is carried out should influence (i.e. Ontology) how we create our understanding of the world we live in (i.e. Our methodological and epistemological choices)" (Aaltonen, 2007, p.58), and the second aspect assumes knowing „the importance of time: especially the use of chronotopes i.e. Places in time, which serve as reflection points for sense-making and strategic decision-making with respect to the situation where they take place as well as conceptual vehicles for managing multiple times and realities" (Aaltonen, 2007, p.58). Thus, for the conceptual building of a chronotop(ic) space, the author uses „the existing unidirectional linear concept of time" (Aaltonen, 2007, p.59) that he submits to some possible conceptual modifications regarding its continuity / discontinuity or the use of different perspectives regarding the relatedness with the future or with the proposed finality for the project that implies a creative process (Aaltonen, 2007, p.59). Thus, this temporal limit (finality) is associated with three types of thinking which he assigns to an efficient managerial process. Within our architectural and urban creation processes we have the same managerial process. The construction of the author's argumentation is very simple and concise as „linear thinking is focused on developing existing practices and businesses. Though it may look further into the future, it is still bounded by existing resources and is involved in allocating them properly. In contrast visionary thinking shifts the perspective, studies the present from the future and by doing that introduces a new dimension into management thinking" (Aaltonen, 2007, p.59). This thinking is a dis-ruptive thinking specific to the real world in which nothing is sure, linear, there is always the chance that something changes without preventing it. Thus, the linear and visional thinking, which are in essence conceptual, are applicable only in the context of general broad thinking, which is disturbing (Aaltonen, 2007, pp.59-60). In this context he describes chronotope as being the unit which „shows the complex trade-offs between the properties of the strategic landscape and the time frame being considered. It is the place where the knots of narratives are tied and untied: a place in time that can be constructed on one level of reality or as a combination of many levels of realities. It is possible to talk about terrestrial or physical worlds, virtual worlds of networks and connectivity, (...) the world of satellites and space platforms, and last but not least the social and psychological worlds that reflect the hearts and minds of people" (Aaltonen, 2007, p.60).

We could add to what has been mentioned before the fact that almost the whole production of architecture and urbanism has a strong symbolic and philosophical load next to the geometric and technical construction principles. From the perspective of our ap-proach this can underline the fact that we are in a post-ideological society in which the cultural landscape becomes subject to the development and at the same time its engine, and the city the a place of progress and opportunity par excellence. We are the witnesses of a post ideological contemporary discourse which gave birth to paradigms such as the cultural industry one, the one of creative industry and last but not least that of urban industry. The permanent change of social life especially in the urban part characterized by both fragmen-tation and centrality, lead to alienation, inequity, departure from nature, negation of the values and redefinition of the way in which we relate with everything that surrounds us. This influences the understanding of certain concepts (and processes) as well as the cultural landscape, but brings at the same time a dramatic change in the plan of individual and collective consciousness. Architecture thus became a tool of interest and power (as in the case of urbanism) and the economic, social and identity crisis the modern societies deal with now threats the ideological fundaments.

Architecture gained thus legitimacy and became a desideratum of development. This is an almost never-ending source which can feed the usual needs of individuals. This is what Francoise Choay affirmed regarding the problematic of the scale tackled in archi-tectural and urbanism planning and the fact that solving this desideratum implies a collec-tive awareness of the society on the actual reality in which „the historical town, the villages and the landscapes are today one and the same ting. Can they be left at the disposal of cul-tural consumption? Isn't it the time to make creations of them again? The still so present European town, even if drastically shattered has to and could be kept and used at the same time as work of art, as social patrimony and as incitement to finding again the ladder and urbanity. Is there still time?" (Choay, 2011, p.130). There is the need to redefine the way in which we relate with the space, and this supposes first of all adopting a philosophical op-tion (Choay, 2011, p.130), dialogical in which architecture (and urbanism) play an essential role. The dialogism of this action consists in the elegant, subtle and solid solving of the way in which "societies will know to rediscover its [architecture's] essence and reorganise education? Will architects learn again the three-dimensional experience of space and art of articulation? Will they find the way of modesty to give the subjects their founding role back?" (Choay, 2011, p.130).

We thus distinguish the necessity of renegotiating the status of architecture and the architect within the actual social frame in order to define in terms which go beyond the ideological spheres the contemporary human condition. This necessity resides from the need of knowledge and deciphering the truth and the reality. Here are also knowledge and deciphering the cultural landscape and at the basis of this process there has to be a solid theoretical foundation, which favours the innovation and development of knowledge if constantly sustained and completed.

The chronotope of human intervention within the studies of cultural landscape is architecture which „ through its preoccupation with aesthetics, codes and the spaces of our everyday lives, has a distinct advantage to affect culture by symbolic means" (Naik and Oldfield, 2010, p.23). We think that this very desiderate gave birth to the initial concept as well as the later studies of the cultural landscape of Carl Ortwin Sauer or of John Brinckerhoff Jackson, in which rural communities played an essential role. But, as it is known, limiting the study of this concept to a unique subject proved to be importunate, so that the study of landscape and cultural landscape became an inter, plural and multi disciplinary one. In a study on the perception and evaluation of the [cultural] landscape Frederick Aalen presents a description of the subjects which are implied in these processes (Aalen, 2004, p.14). Among them he mentions the architecture of landscape, its planning and urban and rural planning. Still, an important and crucial aspect in the construction of the concept of cultural landscape, as it was first described, was the anthropic factor. This factor gathers elements which are chronotopes, bearers of values and symbols which define the space and which are witnesses and at the same time actors of passing of time.

The transformability of the cultural landscape through human intervention leads to the formation of the dialogic character of the system, which is created determined by the landscape, individual and architecture. The three components are simplified for a basic understanding, but these are parts of a major system and they give birth to relations both inside a single chronotope and between two or several different chronotopes. In order to understand this, we come back to the basic work of Mikhail Bakhtin (1981, pp. 252-253). He discusses about the existence both a chronotope of the author and of a chronotope of the reader. For the theoretical foundation of the two chronotope types, the author proposes the construction of two types of reality, the one of the work and the one of human exist-ence. Thus, Bakhtin says, out of the real world chronotope, which serves as a source of representation, the chronotope created of the work is born. (Bakhtin,1981, p.253). In the case of studies about cultural landscape is easy to detach the two worlds, maybe even easi-er than in literary works. Thus, the chronotope of the author is creation in the case that we, architects and urbanists, are interested in. The chronotope of the reader is transformed in our case into a chronotope of human existence through / in the spaces created by the "au-thor" on the different time levels. The chronotope of the cultural landscape is built of at least two chronotopes, the ones mentioned above, each of them being composed of other two types of chronotopes. As an example we can mention the fact that not only the strongly anthropized landscape is cultural landscape. Besides the urban or rural areas, which also are components of the cultural landscape, the latter one can be represented by certain landscapes with strong symbolic load, spaces of collective memory of one or more communities.

According to John Brinckerhoff Jackson, the cultural landscape can be represented by a landscape passed through by a high way, by a chalet in a mountain landscape, a cultivated land, a dam, a surrounding, etc, practically any human intervention in nature can represent cultural landscape. Thus, the subcategories of chronotope existing in the major system of the cultural landscape are almost endless regarding the configuration types.

The chronotope of the cultural landscape, as a result of human intervention presents the essential component of defining the concept that is the social and human aspect. Thus in contrast with the other subjects implied in the study of the cultural landscape, which, as Frederick Aalen showed too, are mainly preoccupied with the historic study as well as with the analysis of already existing data, architecture and urbanism next to all the other subjects creative participant, are those which through the above mentioned characteristic dialogism determine the evolution of the chronotope and are those which define the past, the present and the future of its existence [of the chronotope] in a dynamic spatial-temporal system. Mikhail Bakhtin put forward the existence of a chronotope of the author. According to the Russian philosopher this is both inside the work and in its exterior. (Bakhtin, 1981, pp.254-257). We have to underline and mention the fact that the chronotopic analysis of the cultural landscape, as it is underlined in our paper, represents an interpretation based on a strong scientific argumentation of the concept imported from the literary study.

Thus, the parallelism between the author of a literary work and the author as an "artist" of space (social) represents a construction we will use in order to emphasize once again the fact that the cultural landscape is a vision (as we showed in other previous works). Bakhtin's contribution helps us especially in emphasising one of the main characteristics of the cultural landscape, which we mentioned earlier, that is its transformability. We consider that this characteristic is sustained first of all by the very duality and dialogism of the operative concept, which is the object of our analysis.

The transformability of the cultural landscape is directly subjected to materiality and spirituality of the creative process of space, which as known presents for the creators [among which are architects and urbanists] an enormous reality of the sources of philosophical, artistic, metaphorical, literary inspiration as well as those scientific, but the latter are rarely used. Thus, in contrast to what the study of cultural landscape represented at the beginning of the last century, nowadays we assist at a change of paradigm in many implied subjects especially those which are the object of our profession, architecture and urbanism.

Mikhail Bakhtin' conclusion in the essay about chronotope which were added after almost 40 years (Bakhtin, 1981, p.258), helps in adding what we said above. Thus Bakhtin emphasized the fact that „ the represented world, however realistic and truthful, can never be chronotopically identical with the real world it represents, where the author and creator of the literary work is to be found" (Bakhtin, 1981, p.256). A deeper analysis of what the Russian philosopher affirmed determines the fact that through the creation, construction and planning process our connection with the world is realized through the social-human, artistic and philosophic relations, which architecture and the constructed space determine. In conclusion the creation process, even if it is produced in the real world of the creator and user, at the chronotopic level the two worlds are different. The world created through the vision on reality lead to direct modifica-tions, which are beneficial or even through al-terations of the present world. Both the creator, the reader or the user are actors of the space-temporality of the creative process, which determine chronotopes specific for every world.

We assist at a change of paradigm of the space in which complexity, dynamics, movement and "the becoming of the shape" determine the contemporary evolution and lead to the transformability of the cultural landscape as Mikhail Bakhtin said: „the work and the world represented in it enter the real world and enrich it, and the real world enters the work and its world as part of the process of its creation, as well as part of its subsequent life, in a continual renewing of the work through the creative perception of listeners and readers" (Bakhtin, 1981, p.254) of the reader or the user. Bakhtin continues affirming that this very changing, adaptation and transformation process is chronotopic. Thus „we might even speak of a special creative chronotope inside which this exchange between work and life occurs, and which constitutes the distinctive life of the work" (Bakhtin, 1981, p.254). This special creative chronotope is the cultural landscape which we called earlier in the book ideal chronotope. A chronotope of meaning which determines the spatial-temporality of human creation associated by Mikhail Bakhtin and by a certain semantics (Bakhtin, 1981, p.257).

Can we talk about a semantics of the cultural landscape? As we presented in the previous chapters there is a language of [cultural] landscape. Thus we can deduce the fact that there is a semantics for it. Semantics implies attribution of meaning and signification to certain complex notions which can be derived from simple symbols of the language. Referring to this Bakhtin said that „meanings exist not only in abstract cognition, they exist in artistic thought as well. These artistic meanings are likewise not subject to temporal and spatial determinations. We somehow manage however to endow all phenomena with meaning, that is, we incorporate them not only into the sphere of spatial and temporal existence but also intro a semantic sphere" (Bakhtin, 1981, p.257).

9.3.4 Chronotope and chronotopy at the architectural and urban level

Continuing the argumentation from the previous chapters, in which we showed the status of chronotope of the cultural landscape and the particularization we presented in the chapter above, we thus reduce the analysis to another scale, the one of the architecture object and we continue by presenting in details the subject regarding the urban space and even more detailed the one of cultural landscape.

The definitions given to architecture are as many as different and in order to understand the role of its chronotope we won't appeal to any of those already known. Maybe a short reference to the definition given by the encyclopaedic Romanian dictionary is appropriate. It defines architecture as being „the science and art of organisation and building of the necessary spaces to human life and activity"[149]. Thus, to the purpose of our argumentation we will extend the meaning of this definition. Consequently, chronotopy meaning the intrinsic relation space - time is determined by the representative and social aspects, which architecture attributes to space. First, the chronotope developed as being the way in which historic time is assimilated to space, especially in literary works. As we could observe in the previous chapters by extrapolating the meaning, the result can be of a real help in building our argumentation regarding the study of cultural landscape. We will tackle a series of other themes, which we find in the previous chapters or in the published book in order to be able to analyze them in detail regarding the concept of chronotope. Aspects such as language within the study of cultural landscape, the concept of the town as a book and its discovery as a text or the concept of palimpsest will be the subject of our analysis in the following chapters.

Let us come back to the role of chronotope of architecture. Thus, examples of chronotopes and analytical categories taken out of literary studies, which can represent components of the study of cultural landscape taken even individually could be the following: the road, the inn, the lighthouse, the pub, the caves, the castle, the provincial town, the chalet in the middle of nowhere and many other elements. Let us here remember the series of elements of the historic cultural landscapes such as those mentioned above. Thus these components represent what Bakhtin called minor chronotopes but these are parts of the major chronotopes which are, in our case, architecture and the urban space and on the top of the pyramid the cultural landscape.

Architecture is the one which determined dialogic spaces in the town, which determine connections and dialogues with the landscape. Both modern architecture and the post-modern one create connections with the existent historical buildings being the engine of representative and social determination of the urban space. Thus, through the language of literature, which combines different styles from different historic periods, and which use the existence of the multitude of theories, we obtain an heterogenic approach of space through concretizing some concepts which defined the theory of architecture along the last century. We can thus affirm the fact that the architect/urbanist (the local artist, the creator, the planner) is like a film director who has the task to please both the actors and the spectators by the artful use of several concepts, theories and techniques that are at his disposal in order to create. The creation of the architect and sometimes even more important the one of the urbanist will mould and dictate even more the life of clients, of the users, of the citizens of a particular town/community.

---

[149] The definition is given on-line by the encyclopaedic Romanian dictionary: http://www.webdex.ro/online/dictionar_enciclopedic/arhitectura accessed on 7 February 2012. Our translation.

9.3.5 Chronotope as a basis of cultural production

Reminding the fact that chronotope is a kind of "bringing together" of time and space in the literary work, in the case of architecture the two components are inseparable and form the very frame of life. We have to remind the fact that this whole daily scenography, which is the very product of architecture and urbanism, is the engine of social, economic and cultural development. We can thus affirm that architecture represents the frame of a scene in which the individual is captive and obliged to live his life. This creates spaces of memory, which leads the chronotopic aspect to what professor Augustin Ioan called a theory "on the <<spongy>> character of architecture which absorbs time in order to give it later back" (Ioan şi Mihali, 2009, p.87). Thus, these determine the fact that there are four categories of architecture such as "time before the house", "the time of creation", "time after the creation" "time of dis-creation" (Ioan şi Mihali, 2009, pp.87-97).

The architect Josep Muntanola builds the chronotopic thesis on architecture based on the studies created on the way children understand and relate with the environment they live in. This is essential as "the children know that the places are mediators between nature and culture, between the social and physical dimensions of life and between spatial and temporal qualities of our culture and the social meanings of these <<qualities>>" (Muntanola, 2004, p.5). Thus, the educational role of knowing the processes, relations and elements, which determine and mould the cultural landscape as well as knowing the theories which are at the base of the concept and the way it is understood and adapted, is essential in the acceptance of the existence of cultural landscape as Weltanschauung.

Essential for the profession of urbanist and of architect is to facilitate this knowledge by our own understanding and good application of the concepts, theories and specific practices. We will thus apply to the narrative characteristics of architecture and of the urban space in order to emphasize the importance of their role of chronotopes.

According to Josep Muntanola Thornberg (2004, p.2), the chronotope is the very base of the cultural production considering the fact that everything that is created by man is chronotopic. Thus, architecture is considered as having a chronotopic structure as is binds the social and cultural structures, which determine the space (Muntanola, 2004, p.2). The Spanish architecture presents three chronotopic dimensions of architecture (Muntanola, 2004, p.11) determined by projection (or design), living and the lived space. Each of these dimensions determine relations time-space, which Muntanola considers as being the following: planning determines a mental connection time-space called „psychogenesis"; the built space determines a cosmic connection of time and space called „topogenesis"; and living determines a historical spatial - temporal connection called "sociogenesis". Thus, Josep Muntanola Thornberg (2004, p.3) concludes, the reason why a chronotopic analysis of architecture has not been created yet is the fact that the chronotopic structure of the environment in which we live is studies only through its permanency and not its changes.

Consequently from the point of view of the architect, elements such as the limits, barriers, the limits between the interior and exterior (within the town) or between culture and nature, are important for the chronotopic study (Muntanola, 2004, p.3). Thus, the condition of limit or cog as well as the system produced by these are determined by movement, transition and vision, which creates the chronotopic connection between the mind, landscape and society (Muntanola, 2004, p.3). It is essential to add the fact that the whole society is built on a spatial-temporal structure, in which architecture and urbanism are tehtools of a dialogicconfigurative transformation within which we as individuals are at the same time actors and scripters as Josep Muntanola used to say, we are at the same time masters and slaves (Muntanola, 2004, p.5).

9.3.6 The concept of chronotope and the relation time-space in urbanism and architecture

Following what has been written so far regarding the role of chronotope of architecture we considered it important to analyse the concepts of the relation time-space in architecture of the philosopher Paul Virilio. For this we appealed to the work of John Armitage (2001) in which there are presented a series of interviews with Paul Virilio and discussed the most important concepts used along time in his book. Before we do this, it's worth mentioning a short paper of Thomas-Bernard Kenniff (2011), in which he refers to a series of French radio shows of Virilio, in which he talked about the chronotopic role of architecture. The French philosopher transposes a series of concepts often used in intellectual discourses, such as time and space, in the area of architecture with direct implications on the meaning of the concept of Bakhtin's chronotope. The fact that architects manipulate space is a well-known fact, but the acceptance that there is a difficutly in transposing their projects in (chronotopic) time-space is underlined by Virilio (Kenniff, 2011), this fact showing the lack of temporal evaluation of architecture. As a simple process architecture is registered in time, which can determine its existence as chronotope. But how can this fact be produces? To this effect, Paul Virilio (Kenniff, 2011) emphasises the importance of understanding Bakhtin's chronotope as being the very presence of some spatial-temporal aspects in the project of architecture. In essence we can talk about the intersection of some spatial-temporal sequences which give birth to chronotopes both in the architecture project as well as its creation process.

For Virilio, these chronotopes of the architectural process are those which produce space. To this effect we have to extend the chronotopic study to the architecture practice and school (and urbanism at the same time) in order to understand it, but we will come back to this later in the work, after our argumentation is better developed. Virilio reminds of the fact that in the process of architectural creation a series of individual chronotopes are created of each directly implied entities, such as the one of the client, the architecture, the constructor and the final user. This leads to the different understanding of time and space, which determines the temporal aspect of the architecture practice. In order to conclude, Paul Virilio affirmed that architects are not appropriate to act in time, but in their process they are directly implicated in this very temporal aspect, difficult to count and control, if not even impossible. The inseparability space-time is bound both by the creation process and the architecture object and this implies the existence of the possibility to manipulate time by the urban form and the architecture object.

As we are already used from the important works of Paul Virilio, the problematic state of time and speed as well as the one of virtual space represents the main point of this research ion the last thirty years. Thus we can understand the importance of treating a subject such as the chronotope in architecture. The everyday reality, the fact that we live in a cinematic society of speed and movement, reducing the distances regarding the time to pass through, all these, such as Virilio affirmed (Armitage, 2001, p.17), make us not to get rid of Einstein's theory of relativity, as difficult it is to understand its original form, as we are part of it, we live through it and with it, it is part of our life. Thus according to the French philosopher we all come to interpret the theory of relativity through the lives we live (Armitage, 2001, p.17).

One of the important concepts of Paul Virilio regarding the study of space and time is deterritorialization. He attributes the concept of the alert style of everyday life to the fact that with our program, relation and education (Armitage, 2001, p.17) we get to be captive of time and submitted to space. This fact is associated with deterritorialization, a process which affected each of us. Our rootedness, what Virilio calls here and now in situ, is considered to belong to be past (Armitage, 2001, p.17), as the speed with which everything happens in everyday society topped territoriality, so that reality is unfolded between two dialogical poles which are the geographical (spatial) and temporal, deterritorized frame.

In this context a second important concept is the critic space. Thus, deterritorialization determines the critic situation in which space exists, being constantly suppressed to some changes which lead to the loss of unity (Armitage, 2001, p.24). Space is constantly submitted to simultaneous destructions and reconstructions, what fragmentizes, decomposes and destructurizes it. This is called by Virilio as being "an accident", a constant theme in its theoretic approach.

This critic situation, its deterritorialization and fragmentation lead to another important concept, essential to the chronotopic study, disappearance (Armitage, 2001, p.33). Paul Virilio affirmed that everyday society is characterized next to speed by disappearance, which is compared with the aesthetic of apparition (Armitage, 2001, p.33) of the antique societies. He associates disappearance with the apparition of technique giving as an example the evolution of art and photography devices and techniques as well as the apparition of the cinema. He considers these to be eloquent in order to in order to understand the concept of disappearance which characterizes modern society. Everything has a certain speed, the time associated to certain activities reduced considerably, so that everything became a metaphorical instant (associated with the photographic technique) or a succession of such instantaneous frames. This is how, according to Virilio, the aesthetics of disappearance appears (Armitage, 2001, p.33). The persistency or the pause necessary to assimilate some everyday aspects transformed from a material datum into a mental, cognitive one, so that the existence is due to the disappearance of these actions. This disappearance is applied to things and activities and the speed everything runs with determines a temporal crisis, which characterizes the whole human activity.

In this context Virilio and the concept of oblique function appear (Armitage, 2001, p.53), which denote a certain state of the individual, a special one, which determine him to withstand by creating a instability vision on a stable situation. In order to explain Paul Virilio reminds of two contextual states of this oblique function, one of architecture and the other at the urban level. At the level of architecture Virilio emphasizes the necessity to go beyond orthogonality towards an architecture based on topology as a vision on a third spatial dimension, the oblique one, next to the other two, which he considers to be the vertical and the horizontal. In the urban context this oblique function was considered to be a third urban order. The other two were in Virilio's vision (Armitage, 2001, p.54) the following: the first order was the one of villages and horizontal settlements; the second order was of the vertical mega-structures; the third order was of the obliquity implying that town will be able to develop both linear and topologic by orientated surfaces.

In the interviews we mentioned din the last paragraphs about the important concepts of Paul Virilio it is mentioned an extremely sensitive actual and very important issue in our profession, the political one. In the context of our study and of the argumentation of this work, regarding chronotopy, Virilio's answer to the question whether the oblique in urbanism means political is able to rash extremely delicately this issue of urban spatial-temporal structure of the urban chronotope. Paul Virilio (Armitage, 2001, p.55) answers by emphasizing the fact that there is no simple political but only (geo)political, where "geo" refers to geography and geometry, with direct implications on space. The substrate of this affirmation is extremely simple managing to raise several questions. Virilio explains that in the 60's of the past century he came to the conclusion that in order to intervene in space you have to use power (Armitage, 2001, p.56), which determined him to concentrate his research on the study of time and speed as, in his acceptation, speed means time and political at the same time.

Thus, architecture is dictated by time as much as it is dictated by space, and the very technological and scientific progress lead to a movement change even in the case of buildings. Thus, to face this change, this (geo)political mentioned above in the context of our study, an auto-inclusion of each element into the other is determined. In essence, geometry is included in geography by what Virilio calls strategy, giving the example of Bentham's Panopticon as a central point from which all the others can be controlled, which leads to a transparency of architecture and urbanism which is unidirectional. The strategy, in the case of our profession becomes a procedure, which determines a constant duality between the created object, its location and destination. Virilio classifies this duality as being a process of mobility, by which the town are no longer evergreen, they becoming themselves mobile elements of a world which is continuously moving. Time becomes superior to spatiality. The chronotopy of architecture and urbanism takes place by movement, the movement being the one which transforms space and its perception. Even if the space is "spatially" conceptualized we need a certain scenography to bind everything according to time. Thus, even if the production of architecture is in essence static, it is defined by a certain movement in space (Armitage, 2001, p.61), and speed is the one which dictates the way in which the town is perceived.

According to Virilio in the urban environment a new concept appears, anticipation (Armitage, 2001, p.62). Thus through movement the speed (which represents space and time at the same time) is the one which determines anticipation, which imposes a certain degree of control of the movement action. The transparency we mentioned earlier valid in the case of strategy, of control being unidirectional is valid in this case, too, in which anticipation is possible only through control on movement and its effect. Briefly Paul Virilio talks about a discharge of the horizon (Armitage, 2001, p.63), as in the case of Paris happened or the glass facades so much used in modern architecture.

These procedures give birth to spaces which can lead to a certain control of the movement and in essence of time, as speed constantly leads to a change of perception and in the best case this can be anticipated. We talk here about a demultiplication of time as Paul Virilio affirmed. Being helped by the definition of demultiplication founding the Romanian dictionary, transforming a rotation movement into another slower rotation movement with the help of a gearing[150], we can more easily adapt the meaning of the statement above to the urbanistic architectural study. The result of the spatial configuration both architectural and urbanistically, due to the speed with which the human activity is unfolded leads to the necessity of a temporal approach regarding the experience of space.

[150] Definition found online at, www.dexonline.ro. Site accessed on de 3 March 2012. http://dexonline.ro/definitie/demultiplicare/paradigma.

9.3.7 Movement: main process in constructing an urban chronotopy

Consequently, in order to build the argumentation regarding the temporal demultiplication, we pass over from Virilio's theories to the analysis of Wolfgang Meisenheimer (2007) over space. The well-known professor, supporter of the oriental philosophy in architecture, had a great impact on the way in which space and time are studied through movement in space. The volume of works we will refer to brings together the most important ideas regarding the way in which space-temporality has to be seen in the process of urbanistic architectural creation.

Thus professor Meisenheimer said that, in order to be able to create a proper environment for everyday life, we have to base in the creative process on a valid theory of spatial-temporal models (Meisenheimer, 2007, p.169). To this extend the presentation of the chronotopical theory of architecture and urbanism can be easily built on the influence of professor Meisenheimer's concepts. In his books he presents a series of space catalogings known at the conceptual level by architects and urbanists, but the major contribution for the study of the chronotopy which we initiate has the concept of passageway[151], (or connection, road), which by use leads to understanding the process of demultiplication of time in the urban environment. Thus, at the level of all known spaces in the urban public environment the activity types are essential for defining the spatial-temporal model. These activities are grouped in systems, which determine relations. At the public level all these systems and relations between them are determined by movement, which, according to the professor determines the public form, that Gestalt in German as being a product of time and of movement. In this process of public formation the architectural construction is a kind of guide, an element of orientation, focalization which leads to the creation of the metal map of space (Meisenheimer, 2007, p.106). The mental map determines the spatial imagination of the individual, which together define and are influenced by experience.

This urban chronotopic experience is determined by a series of elements or

[151] Here we have to add some things in order to be sure that the concept is understood. Through the term passageway, we understand a way, a street a walking way, etc. The use of the concept as Meisenheimer uses is essential in order to keep its initial meaning and its authenticity. Metaphorically the term passageway is useful in order to understand the process of transition (from the verb to pass) of transformation in the urban area due to movement. The whole argumentation regarding the chronotopy of architecture and urbanism is built on the process of movement, the one that combines both spatial and temporal experience of a space. The movement transferred into speed, an essential characteristic of contemporary life is at the urban level an important component of experience. This is in an antithetic dialogue with the process of contemplation used in the case of perceiving the landscape. We find this dialogism important in order to show the structural and systemic differences which occur in the process of studying the cultural landscape the two mentioned levels.

analytical categories we named at the beginning of our studying this concept. Besides those taken out from the literary works, at the urban level we have to extend the list reminding the important focal points such as the church, the mayor's house, the market next to the castle, inn or pub, as being the most important and which we have often met since the apparition of the first towns. The railway station is an important focal point, which appeared once with the development of science and technique and the industrialization, having a more recent history. Still, public life is different from town to town around these markers. Between these markers, major chronotopes in essence, there is an intimate space of markers and little signs, singular and personal, sometimes individual, which build the identity of each individual or of a small community. Their totality, according to professor Meisenheimer, represents the foundation of everyday life (Meisenheimer, 2007, p.104). All these markers have their time given by the form and style of each of them (spatial elements temporally determined) which form what the teacher used to call "temporal perspective" (Meisenheimer, 2007, p.106), the equivalent of the spatial one but determined by "temporal distance" (Meisenheimer, 2007, p.106) of each element given by the history period it appeared.

These urban markers are connected by passageways, ways, roads, which, as we will later see, seem to be the most important chronotopes and maybe the less studied ones within the cultural landscape. But up to the study of the chronotopy of the road (which is part of the chronotopy of architecture and urbanism), the analysis of passingways is essential in order to understand the spatial-temporal models. Thus, the spaces defined by the local, urban markers are structured and related by the transition, temporary or moment spaces. Professor Meisenheimer (2007, p.108) calls temporary spaces also the fast-food stalls, the removable markets, the camps or the little festivals. These are spaces which appear and disappear, sometimes after a certain model other times accidentally. Elements of architecture show movement by their nature, and these spaces are spaces which are dictated by movement and speed, and time is assimilated in space as a connex phenomenon, which determines a virtual reality (Meisenheimer, 2007, p.124). This is as the passageway implies the existence of two distinct and important moments: the beginning and the end and everything there is between these two represents time and the movement along this way is determined by the spatial forms (Meisenheimer, 2007, p.124).

These entries and exists are closely connected to what Paul Virilio said regarding anticipation and disappearance. By the access to the passageway anticipation of what will be discovered and experimented is produced and by passing through the passageway through movement and speed disappearance of what is left behind appears which implies an attachment of memory. Thus, the architectural and urban structure of a space has to be thought not just through spatiality but also connected to time as movement in space appeals to memory and experience. It is easily understood the fact that the spatial dimension is subjected to the temporal one and for professor Meisenheimer the structural forms of urban passageways are part of a temporal concept rather than a spatial one. Mentioning Paul Virilio at the beginning of the paragraph we have to emphasize the fact that what the French author said about transparency in architecture and urban-ism was continued by professor Meisenehimer in a slightly different way. He is preoccupied by the way in which transparent facades transform the urban space into a metaphor of interior space and the variety of structures, textures, materials and colours of facades make them transform into "interspatial membranes" (Meisenheimer, 2007, p.150).

Thus the facades become spaces of the transition itself, areas of anticipation, announce and memory. And all these passages or passageways produce a real inflation in the urban space and become instrument of a choreography which determines our speed in space and automatically determine time, too. The urban chronotopy is thus composed of the spatial conformation of the place and the principles of the apparition and experience, which are transferred on a temporal structure of the daily. This very issue of spatial-temporal simultaneousness, which the concept of chronotope imposes, was discussed about the Romanian philosopher Alexandru Dragomir. He identifies the issue as being the fact that time cannot be expressed otherwise than by space, which contrary to the first one is static (Dragomir, 2008, p.70). Consequently the Romanian philosopher considers it opportune the conceptualization of the fact that "space is where there is no time" and "time is where there is no space" as being a defying method of the first by the later one. Thus, in order to obtain the chronotope, what leads to transformation, what makes out of the cultural landscape the product of some social, historic, geographical, economical and political conditions, which never are neutral towards the relation system, which are based on the human intervention.

In Meisenheimer's argumentation we find a reference to the mentioned fictional part reminded by Bakhtin about the chronotope, emphasized through movement which defines the everyday life of our century and which determines reality that unfolds around well defined spaces which produce a spatial-temporal dependency of human activity both real and imagined. To this effect, Meisenheimer affirmed that fact that these everyday activities appeal to imagination (Meisenheimer, 2007, p.174) and memory and experience creating the drama of the transition process in the urban environment (Meisenheimer, 2007, p.181). The theme of urban drama is to be found at Lefebvre who considers that this phenomenon of drama is suppressed to a contradictory dialectics which can determine the spatial-temporal transformation of the urban reality (Lefebvre, 2003, pp. 124-125), enumerating the processes which determine this fact and among which it is worth mentioning that of the dominant activity which produces the spatial and temporal change. This activity mainly of the urban space can thus determine either "the disappearance" of time through space, or its transformation in space (Meisenheimer, 2007, pp.245-441). These actions imply a rupture, which according to Lefebvre (2003, pp.122-124) is determined by the spatial-temporal change and the apparition of a fortm of knowledge at the urban level based on a specific language, a (political) discourse, argumentation and reasoning, analysis and capacity to act.

All these mentioned actions and concepts in the previous paragraphs, which appeared in the theoretical discourses of Paul Virilio, Wolfgang Meisenheimer or Henri Lefebvre, are specific of an urban chronotopy and of that intrinsic bound space-time mentioned by Mikhail Bakhtin. For Lefebvre, the connection space-time at the urban level is represented by the centrality which is specific of the city, which is correlated with the dialectic movement from the urban area and the fact that any time a point from space can become center (Lefebvre, 2003, p.116). This dialectics of movement is considered cyclic and confuse, which determines the urban space to become in this order "conceived, perceived and unfolded" (Lefebvre, 2003, p.117). Movement is, as we can notice, the main element of chronotopy which we want to build at the urban and architectural level. Thus in order to extend our argumentation about the concept we will continue referring about the chronotopy of the road or the street as an element of spatial-temporal transition.

9.3.8 Chronotope and chronotopy at the level of the street

We will continue by going deeper into the study of chronotopy towards an field that we consider weekly developed if not even absent in the studies of cultural landscape. We have to make some supplementary explanations about the study of the previous chapter. Reminding the fact that in the previous chapters we developed a full theory regarding the conscience and chronotopy, the role of this latter underchapter is to bring clear examples about what we understand through chronotopy at the level of cultural landscape. Thus we proceeded to a "descentralization" of the concept what we want to be an explanation of the chronotopy of architecture, urbanism and landscape in the studies of cultural landscape. The purpose of this study is to theoretically determine the understanding of spatial-temporality in the study of cultural landscape. To this effect from the analysis of spatial-temporality, of architecture and urbanism through the concepts of the four theoreticians studies in the previous chapters, we came to the conclusion according to which space-time in the urban frame represents movement and centralization at the same time, disappearance and anticipation and last but not least transition. Thus in analogy with the described passages by professor Meisenheimer we consider the chronotopical study of the street, the parlour, the road, the passage, or the passageway as being proper.

Reminding about what professor Alexandru Sandu considers about the concept of cultural landscape, that it is tightly connected to modern living and as a result by a certain mentality of living (Sandu, 2007b, p.27), this has a great influence in the context of spatial-temporal determination leading the meaning to the urban life, and this draws our attention towards the street, at the urban level, and at the rural/spatial level towards the alley, path, road or footway.

The everyday takes place in the street, the urban life takes place in the street. On this almost neverending topic for the theories of architecture and urbanism there have been published very many works but we will continue by referring to only some of them which we consider emblematic for the creation of the concept of chronotope.

We will start the discourse about the issue of street chronotopy by referring to the very process of contemplation known from the study of landscape. This process at the level of landscape is a static process but in order to reach this state of contemplation we use movement. Thus the visual part together with the olfactive and hearing one is the most frequent used, at the level of the landscape being developed by the need to stand at a topological level which should facilitate the vast vision on the natural/anthropic area. The panoramic frame is thus every time close assimilated to the process of contemplation. The need to pass through space in order to reach the point of maximum visibility makes that on the way the importance given to the area passed through is minimal. Let us remember what it meant for each of us passing through a forest only to reach a broad space which could permit a good and fascinating panoramic view on the landscape. Thus the movement on the way distributes the attention only towards the fulfilment of the final purpose; there are too few elements or events which can distract us from our movement process. Thus this passing through is essential as it implies transition. This transition is determined by the speed, which is determined by fulfilment of a purpose, the very purpose of space travel.

At the urban level the movement along the streets resembles the one in nature which determines the need to have also broad spaces, which could permit the change of speed and repositioning from the transitioned space. The street or the road/path is thus the one which opens the space (Bollnow, 2011, pp.94-95). At the same time this has three important functions: the informative one, the symbolic one and the ludic one (Lefebvre, 2003, p.18) which determines its chronotopic function by what we will demonstrate in this chapter. But continuing the above mentioned idea we have to say that the city does not have to be suppressed to contemplation as in nature. This is clearly argues by Bogdan Ghiu who says that "the city is no picture, a museum or a film, an exhibition or a concert. The city can not be contemplated. The act of contemplating a city, of <<visiting>> it, of watching or focalizing is a trap. Because, through the very sight, we are caught and the city is stolen from us, used against us. We assist today at a neutralisation of the city under the form of a communicative channel or support, at its transformation into background: panel and simese" (Bogdan Ghiu quote in Ioan and Mihali, 2009, p.13). This almost radical thinking is, we can surely admit, extremely legitimate. We consider the neutralization of the city mentioned by Bogdan Ghiu to be in close relation with what Otto Friedrich Bollnow considered to be objective space, which became system of relations,

(Bollnow, 2011, p.99) which is the city. In this context the urban frame is suppressed to the street and both the individual and the landscape are suppressed to this creation born from the belly of a rational system of power (Bollnow, 2011, p.97).

According to the German physicist and philosopher Otto Friedrich Bollnow, the road leads towards a homogenisation of space, which obtains its whole power by the created net and which is manifested both locally and regionally and nationally determining the control degree of the central power (Bollnow, 2011, p.97). This net produces power but is at the same time produced by power being like the real or blood net of the human body. Thus the space is opened by the road and is at the same time (re)created by (infra)structure of roads (Bollnow, 2011, p.99). This space of objectivity, which is the road according to Bollnow determines an indifference both on the individual and on the landscape. At the level of the second one the indifference is both dual and dialogical: the road becomes indifferent towards the landscape, the walker (the case of the driver who transforms the journey into a drive of the passed time and space (he sometimes becomes indifferent towards the landscape and the landscape is indifferent towards the road. Even if it loses its character and individuality (Bollnow, 2011, p.99) the indifference of the landscape is translated by its continuous insistence towards the human interventions.

Otto Friedrich Bollnow is the one who brings out the problem of the loss of a spatial dimension by passing the road. Thus, being already determined the symbolic meaning f the road by passing through space, through speed, the human nature is changed at the same time (Bollnow, 2011, p.103) once with losing its spatial width. This dimension is lost through the transformation of space, movement gives birth to a specific bound between the experimented space and the state of mind of the individual and the speed is the one which produces the transformation (from pedestrians to drivers) (Bollnow, 2011, p.103).

The loss of a dimension can be assimilated through the reduction process determined by Henri Lefebvre (2003, p.20), by which the street is transformed into passageway as professor Meisenheimer argued. At the urban level the street is the symbol of human interaction (Lefebvre, 2003, p.18), element of cohesion without which segregation would be produced (Lefebvre, 2003, p.18). Despite this the street is a place of disorder, of play and study, a space of parade and show in which the individual is transformed into show for the others (Lefebvre, 2003, pp.19-20). This procedure is due to the problem of "visibility – to see and be seen in the public space" (Ioan and Mihali, 2009, p.44) as "the public space appears when except of me there is somebody else on a given territory and thus there is the possibility not only to watch around myself but also be watched" (Ioan and Mihali, 2009, p.43). Being a transit space, a passage-way, it is a space of communication and exposure and Henri Lefebvre considers it as being a time regulator, a normative element. The street norms the time transforming it into what he called "time of goods" (Lefebvre, 2003, p.20), being suppressed of a system based on economic profit. Thus the above mentioned transition is described by Lefebvre as being "a necessary one between the work force, planned free time and living as a consumption place" (Lefebvre, 2003, p.20).

The condition of street normative is given by the fact that it determines and owns at the same time power as symbol of human interactions on space. Thus it is personified, metaphorically speaking, determining an anarchic movement of the particular and individual interests. The street hyperbolizes the vision of an individualist amount preoccupied with the own thoughts and, as we emphasized earlier, the own interests. The chronotopy of the road (which is often confused with the street) is given by the way in which through movement and transition are correlated the particular / individual, real or virtual worlds.

Thus chronotopy defines the fundamental coordinates of the world: chronos – the movement coordinate and topos – the being coordinate and "the city is the place of fusion between space and time in the sense of passing, a fusion out of which neither time nor space get out the same as they got in. The city has the power (…) to defeat space and time, but not like the village, which makes them eternal and absolute but rather misusing them by relativizing them and domesticating them and suppressing them one to another" (Ioan and Mihali, 2009, p.11). This misuse is due to the temporal structure of the contemporary where the dualism time-space exists only to the purpose of a change.

In the city the street is that fragmented structure which has in its composition a multitude of meanings and symbols and which determines the space we live in to be a mediating one (Muntanola, 2004, p.4). By this term we understand the way in which our experience in the urban area is not singular but every lived and experienced thing is always in relation with the vicinity and the events which take place around. Thus the mediation happens between the mobile and fix elements of the city, the people with their activities in relation with the physical structure of space. All these elements are characterized by permanent successive transformations, which make the chronotopic study a difficult process reminding us what Josep Muntanola (2004) said about the chronotopy of architecture.

The streets are catalysers of (urban) life and experience representing a passageway and happening/passing (Ioan and Mihali, 2009). Thus by the transition from an industrial society to a society of consumption the way in which we relate to the city and the street changed. The happening and passing are components of the street dialogism and determine its chronotopy. Both happening and passing imply movement on both spatial and temporal level which determines defining of a meaning and understanding the processes which are produced at the level of the street. We talk about the narrative characteristic of space in a spatial-temporal continuum and narrativity through the concept of chronotope imported from Mikhail Bakhtin, is he process which determines the meaning reminding us about what we argued in the previous chapters, that the cultural landscape is a chronotope of meaning.

In order to strengthen our argumentation regarding the study of cultural landscape and the importance of chronotopy which we have extended so far through the study of architecture and urbanism it is important to come back to the theoretical construction of the concept, Thus, David Crouch argues regarding the way in which the street is understood and represented the fact that individuals associate the place they live in with the streets around in the way that their relation with the neighbour streets is created by using memory, happenings and feelings, individuals and inhabitants of the neighbourhood, stores, locals and close institutions. Briefly, persons and places which are important to them and which moulded their life in those spaces (Crouch, 1998, p.160). At the same time Crouch affirms that streets assure and offer an imaginative force (Crouch, 1998, p.158). Here resides the dialogical power of the street in the imaginative force which transposes its spatial-temporality in what we earlier mentioned to be its fragmented structure formed of different meanings and symbols. This fragmentation makes the streets stay "places of desire and dread, pleasure and pain, fantasy and fear, their history is full of such paradoxes and tensions" (Jackson, 1998, p.173). At the same time the chronotopy of streets is given by its characteristic to create boundaries "the boundaries between what is real and what is virtual or imaginary; between what is relevant for the meaning of the situation and what is not relevant; between the various phases and sites of the narrative; and also the social boundary between narrators and listeners/watchers, who tells the story and who listens to it" (Lorino, 2010, p.8). To these we add the boundaries / limits which Josep Muntanola talked about, which are manifested at the level of the urban and architectural chronotopy both socially and mentally.

In the following paragraphs we will present what Anne Mikoleit and Moritz Puerckhauer considered to be lessons for understanding the city (Mikoleit and Puerckhauer, 2011). We used their names in the title given to this chapter not in order to make a review of the book published by the two but in order to underline the fact that their "urban code" represents an important source of ideas, concepts and examples for a researcher in the field of cultural landscape especially urban. Our whole argumentation of this paper and of the previous ones tried to determine the intrinsic existence of spatial-temporality of the studies of cultural landscape by the analysis of some distinct components which more or less missed from the well-known studies of cultural landscape.

Thus, in the book *Urban code. 100 Lessons for Understanding the City*, Anne Mikoleit and Moritz Puerckhauer create a trial to deconstruct the urban life of Soho – and to construct of that life an urban code" (Mikoleit and Puerckhauer, 2011, p.108). A part of the so-called lessons which the two offer together with known and lived truths by each of us, no matter of the city they come from becoming real axioms of the urban contemporary life. We will present some of these which we consider important for the study of the cultural landscape and we will enrich their meaning with new theories and meanings.

In virtue of those presented at the beginning of the chapter we will start with lesson 44 as it appears in the above mentioned book. In English this lesson is called: the day to day happens on the street, (Mikoleit and Puerckhauer, 2011, p.53). It is an axiomatic truth known by almost everyone being a fact that could seem weird to be studied when giving the impression of banality. The authors' description will illuminate and clarify what is already known determining the fact that what happens on the street can transform a normal thing into something special[152] (Mikoleit and Puerckhauer, 2011, p.53). In other words the everyday life on the streets is a showcase of our actions. And as any showcase as in the case of market and consumer economy it transforms a normal product into a special one, the best to be bought / sold. Thus, the showcase of our actions, the everyday life on the streets, transposes the way in which we relate to the town through familiarity (Mikoleit and Puerckhauer, 2011, p.53). This familiarity is given y a certain culture of the public space and the everyday which has to be received in any individual one way or the other. But "as the city and its everyday unfold us only in fragments, in splints and stubs impossible to be bound into a theory because our everyday existences are split between the experiences which no Idea cannot embrace, any word about the town and its everyday can be nothing else but fragmentary" (Ioan and Mihali, 2009, p.9). Fragmentary which was mentioned by Peter Jackson (1998) or Paul Virilio (Armitage, 2001), makes it at the level of the fragmented street of our both private and public existence be bound and related and the street being the intrinsic element of spatial-temporal connection at the urban level.

The reverse side of lesson 44 is presented by the one with the number 86 which the authors called: the counterpart of the urban everyday is the urban park (Mikoleit and Puerckhauer, 2011, p.97). Translated into Romanian the equivalent of the urban everyday is the urban park. It is easy to understand that the authors of the book referred to Central Park, the well-known urban park in New York – but this axiom is valid for any other park, either small or large (as the one in Manhatten) of any other city. Being associated o the park we always find the process of contemplation reminded earlier in our work and so vehemently and justified annulled for the city by Bogdan Ghiu. This park is described as „ a place where the concerns of everyday urban life can be released, in exchange for the stillness of the present moment" (Mikoleit şi Puerckhauer, 2011, p.97). But for our argumentation it is not so important to remember the park as a place of relieving, relaxation, introspection and meditation or communication with the nature within the urban frame, but it is the way in which this contemplation state can be obtained even marginally taking into account what Bogdan Ghiu argued at the urban level.

---

[152] Original quotation: „(...) the everyday takes a certain form in the streetscape that makes the usual into something special. An abundance of private everyday events converges in the public realm, lending it a sort of sense and familiarity".

In this connection we will appeal to the lessons number 88: good walkways have a good range of destinations along them and 89: destinations are more attractive when they invite a stay (Mikoleit and Puerckhauer, 2011, pp.98-99). In the context of those mentioned in the previous chapters, of movement and speed, which determine the urban life, these two lessons are, among any others in the book, those which invite us to a more attentive observation of the smallest details of everyday life. The authors refer to a simple passing through the street as passer-by being preoccupied with our individual problems and concentrated on our destination. Thus we become victims of speed, being captivated by our movement which is perturbed only by the apparition of a symbol, a marker of the territory we pass through. As Lefebvre said „the street became a display, a corridor flanked by stores of various kinds" (Lefebvre, 2003, p.19). Thus everything becomes "a substitution of the <<poetic living>> with the <<passing living>> of a metaphor with another one. I could say that the emphasise is moved from <<living>> to passing. Meaning that: in the accelerated microphysics of a urbanity at times seducing than suffocating, living becomes a secondary bet by its <<dislocation>> by its continuous striation with thousands of forces it cannot escape from, leaving thus place to passing as the only desperate chance of each, more and more anonymous and hurried, to meet himself in the corner of the street, at a semaphore, in a car, on a parking place, in a store" Ciprian Mihali said (Ioan and Mihali, 2009, pp.11-12). These places for having fun/passing are to be found (again) at the level of the street and are connection elements without which „ urban life would not exist, leaving only separation, a forced and fixed segregation" (Lefebvre, 2003, p.18).

The last two lessons of the book offer information about the limit concepts and transition, two terms that we analyse in this paper. Thus in the lesson no. 99 it is mentioned the fact that "the main roads are borders" and in the lesson no. 100 the fact that "crevice creates friction" (Mikoleit and Puerckhauer, 2011, pp. 107-108). It is known the fact that due to the heavy traffic the main streets are spaces difficult to pass, the only places to pass are the zebra crossings. These represent a kind of passing places (of the border), transition area by which a territory is left behind and another one is discovered. The authors say that in the case of New York „ streets are an outstanding example of extraordinary liveliness, which has developed in precisely this border zone between two adjoining areas" (Mikoleit and Puerckhauer, 2011, p.107). In this context of crossing limits and transition, Ciprian Mihal and Augustin Ioan talk about a possible transgression, an invasion of limits, especially because "in everyday culture of our late modernity transgression is not an exceptional event reestablishing a meaning altered by the ageing of time and confirming by the act of violation the gathering force of limit" (Ioan and Mihali, 2009, p.10).

But about these limits, the authors misuse in the last lesson of the book the meaning towards fracture or rupture, in order to be better understood in the urban context of transition. Thus „ if we speak about fractures as the means of connection between two places, times, or things, an allegorical parallel to terms like decay, rupture, and caesura inevitably come to our mind" (Mikoleit and Puerckhauer, 2011, p.107). Thus at the end of this little selection of urban axioms, the authors make a short pleading on transition and the way in which these passing points, limits and ruptures determine within projection and urban planning the very "source of a new start" (Mikoleit şi Puerckhauer, 2011, p.108).

In the context of our chronotopic study, within this last almost dogmatic lesson, we find the argumentation regarding what urban chronotopy is in essence being either of street or of architecture: „Indeed, the meaning of breaking points is ambivalent. (…). If one considers breaking points exclusively in terms of time, they appear exactly where one movement ends and another begins. To put it another way, they are the temporal overlaps of both of these movements: the timeframe in which the old no longer counts and the new has not yet found its expression. In these moments of complete lack of orientation, the friction between old and new creates an extraordinary potential for creativity, culminating in a plurality of processes and activities, which thus initiate the change to a new beginning" (Mikoleit and Puerckhauer, 2011, p.108).

9.3.9 Conclusion on the concepts of chronotope and chronotopy in the study of cultural landscape

That semiotic frame mentioned by Kai Hakkarainen is assimilated by William Mitchell (2008, pp. Vii-xvii) is one of artefacts. But in order to understand this world of artefacts, the creative process is the main element of the process of assimilation of the meaning. In order to understand this we have to remember a series of concepts which we used during our long discourse regarding chronotopy and which we will tackle again as a conclusion in this chapter. Before this, we have to understand the fact that this space-temporality is an important element in the study of cultural landscape. This not only because very often the study of cultural landscape is understood as being the historical study, but because of what we mentioned earlier at Josep Muntanola, who said regarding the impossibility to realize the chronotopic study of architecture that this is due to the importance given to permanency and not to the changes arising in the environment. But this impossibility refers not only to the study of architecture but can be applied to the whole study of cultural landscape. In this kind of studies we need to accept that the whole concept appeared in 1908 in the German geography school, then made public in the international academy world by Carl Ortwin Sauer and the cultural geography school from Berckley suffered mutations during a century which have to be shown in the theory built around it.

From the very beginning of the chonotopic analysis on cultural landscape we grounded on the definition given by Mikhail Bakhtin to this concept as being "the intrinsic connection of spatial and temporal relations, which are artistically expressed in literature" (Bakhtin, 1981, p.84). We emphasized the fact that the two categories of time and space are inseparable which is due to the chronotopical construction. At the same time the two categories represent social and historic components which unify for us the concept of cultural landscape, but we have to respect the fact that every epoch had a certain conception and vision on these categories (Brandist, 2002, p. 123). Thus these have always been suppressed to the different paradigms, which governed along time the sphere of theory. But something stayed constant within the different studies, the fact that space-temporality is suppressed to a special dialogism which in Bakhtin's study was subscripted to a relation of differentiation. Out of the man nature/culture are these relations determined and this, as we have shown so far, is transposed in a characteristic dialogism. Thus, we have developed a theory of the dialogism of cultural landscape considering everything a construction of a dialogue characterized by asymmetrical simultaneousness and duality. Within this theory we came to the analysis of the philosophical concept of suppression at the level of the dialogue worn by the cultural landscape which, we considered, makes part of a discourse of several dual entities with which we operated during the whole thesis.

The extension of the notion of time with which we operated during the chapter implied visiting some authors again and some concepts analysed within other chapters from the beginning, but only in order to strengthen the argumentation on chronotopy, and this fact only to determine the way in which cultural landscape is a concept which is used by this very spatial-temporal inseparability which chronotopy implies. This fact completed with the analysis on the social dualism of the cultural landscape lead to the thesis according to which the dialogism of cultural landscape lead to a de-contextualization of space in relation with time. At the same time we came to define an essential characteristic of the cultural landscape as an ideal chronotope and social dualism, it being, as earlier mentioned, the temporal representation of the sequence of interactions and social relations determined by the cultural processes. Thus the theory of the chronotopy of human intervention came into being, though which we presented the analysis at the level of architecture and urban studies, in order to determine a frame of the study of chronotopy which should be specific to our profession. We added to this frame the analysis of architecture and urbanism in order to prove the importance of using the concept in building a valid theory of the cultural landscape.

Consequently using the concept of chronotope in the study of cultural landscape we consider to be a correct and useful way to develop the theory which is around the operative concept. The more the vision of Josep Muntanola is applied, that we mentioned above, by which it is shown the necessity of studying the space we live in, both at the architectural, urban or cultural landscape level by transformations and not by its permanence (Muntanola, 2004, p.3). Extending this affirmation it will be easy for us to observe its credibility. In the way it is understood the study of cultural landscape especially in our country we can distinguish the permanency under which it is studied. It is talked about habits, history and patrimony. These elements we surely cannot avoid in our study and have to be considered not as variables of the analysis but more as some constants. The way in which we analyse the transformations is a very different aspect and this fact can be observed by the way that most of the times this part lacks. Transformability means time, evolution and change. The most important aspect in the study of cultural landscape, we could argue.

10 Final conclusions to the study of the concept of cultural landscape

10.1 The importance of the study and the knowledge of the theory of the concept of cultural landscape

As demonstrated throughout our thesis, being familiar with such a permissive and complex concept such as the cultural landscape represents, from our perspective, understanding all the mechanisms that generated the creation of the concept and its present use in more and more disciplines. The study on the cultural landscape can be translated as a study of changes, of something we may call a history of the future. But why a history of the future when the study on cultural landscape as presented in the paper is mainly related to the past? The answer follows: as architects and urbanists, our role is to produce spaces, to transpose a series of feelings specific to the spaces created. We are creators not only of the past, but also of the future. In the same range of ideas, the expression of history of the future gains some shape.

Referring to the cultural landscape as to a process represents a transition which involves a time and a space dimension in its actions. This transition consists in a series of planned actions, at least for the 21st century. But considering the cultural landscape as a physical and social reality, that mirror of the society we mentioned in the paper, we must also understand it as the everyday (or daily). For this particular reason we undertook the majority of our actions. The cultural landscape is the sum of different realities but particular to every place which came to existence as a result of the cultural influences brought by the human being. At the same time, these should not be separated from the natural where they originated, i.e. the landscape itself. The constant parallel between nature and culture, a dualism that governs over the studied concept, became an antithesis which, over the years, amplified ambiguity and the complexity of our concept.

We can thus take into consideration the need to find new approaches in the study of the concept of cultural landscape, with priority in the Romanian schools of architecture and urbanism. Below we can observe a brief analysis of the most important disciplines involved in the study of cultural landscape. Settling this analysis on an equally brief time scale we can notify that, given the nature of our profession, our involvement in the study of cultural landscape should be (at least) total. That is to say, our study ought to be detailed and to be based on a research as complete as possible because good practice is always the result of well-built theory.

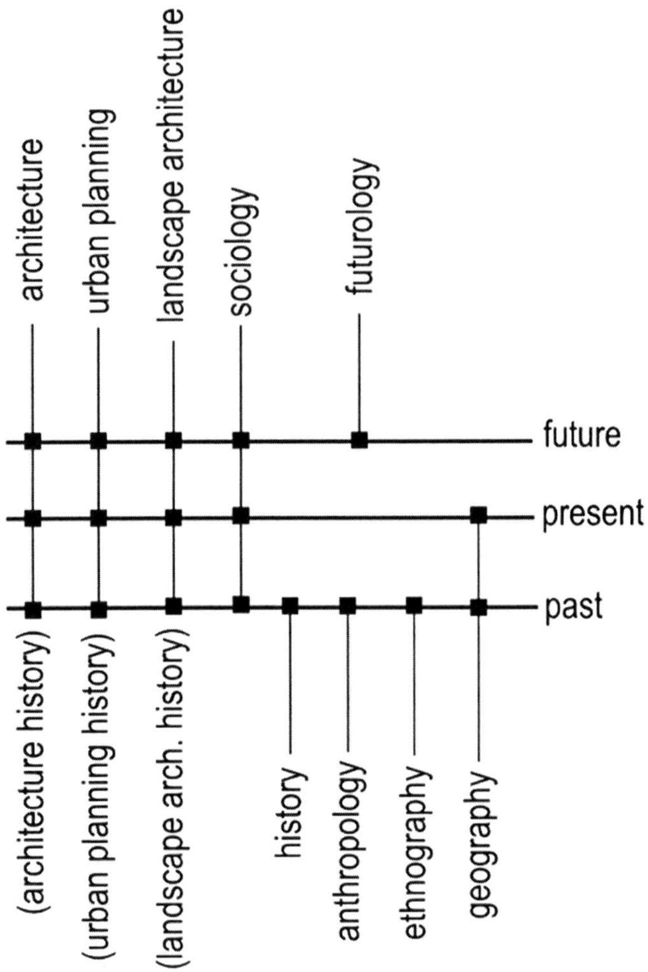

Figure 36. Basic analysis of the disciplines involved in the study of cultural landscape. Author's contribution.

In order to understand the concept and to operate with it we must first of all know the theory that generates it, its evolution and the influences it gained from other disciplines. The explanation consists in the necessity of being witnesses to those transitions we mentioned earlier. And our role as architects, urbanists and landscapers in all this process is sealed ever since the first phases of assimilation of the defining theories for the complexity of our profession. And amongst these we consider that the study of the concept of cultural landscape must occupy an important place in our academic and university training.

In order to understand the concept of cultural landscape we believe it is necessary to approach, at least at a theoretical level, two important aspects related to its existence both as a process and also as an individual element of the everyday. These aspects are innovation and the importance of inventing new elements, techniques, processes, technologies, on the one hand, and tradition with all it brings about in terms of cultural nature which lead to the founding of certain communities, societies, products or processes that resisted in time, on the other. In this direction, the matter in relatively complex and we may ask ourselves if perhaps these two important aspects, innovation and tradition, generate new contradictions in the meaning of cultural landscape. Denise Scott Brown (2010, pp.5-14) believed that these elements are part of some cultural paradoxes. Thus, the study and the understanding of the concept must be related to a detailed understanding of the social changes and the complexity of this transition process. Just like all other change resulted from a transition, the one produced by the cultural landscape brings along with it the knowledge of the entire process. We therefore bring into discussion, besides innovation and tradition, rein-vention because, as Denise Scott Brown (2010, p.13) considered, "the common influence was social change. Shifts in American society spelled a shifting of sensibilities among perceptive scientist and artists. Changing sensibilities induced changing perceptions. These in turn called for a reassessment of tenets and philosophies, particularly in disciplines concerned with urban life". Brown also referred to the American scientist J. B. Jackson's writing, who as we already know, was one of the most preoccupied authors in the study of cultural landscape as an essential element of the everyday.

We discussed about the definition of the concept at the end of the chapter. We were drawing out the argumentation based on the concept as a metaphor and palimpsest taking into account the facts presented throughout the entire chapter. We constantly mention the cultural landscape as a process and an element of the everyday. In order to preserve the connection with what was mentioned in the above mentioned chapter, we must list the relations between the terms presented in this paragraph. We nominated the concept as metaphor and in this range of thoughts, Lucian Blaga (2011, pp.348-349) considered that human creation does not possess a meaning through itself; its meaning resides in its substance which be-comes a metaphor: "as opposed to the substance of real objects from the sensitive world, the substance of creations does not possess a meaning and a role in itself: here the sub-stance seems to replace something else all the time; here the substance is a precipitate which implies a transfer and a list of terms that belong to certain regions or domains. Substance thus receives a so-called «metaphorical» aspect". This easily makes us think about the terms of innovation, tradition and reinvention because they are all closely related with the human specific character and the way the meaning of all that surrounds us is created through cultural processes. Annamaria Lossi (2010, p.213) stated that metaphors always express a different type of reality , which in the case of cultural landscape is constantly (re)created either through preserving traditions, or through innovation or (re)invention; they all create the everyday of our lives. The latter is a worldview (Weltanschauung) and at the same time a mirror of the society which produces and consumes it.

The daily possesses a particular narrative character we mentioned in the thesis with-in the cronotopic analysis of the cultural landscape, through the palimpsest-metaphor relation. Speaking of palimpsest, we consider that it has already been understood how we referred to the term within the cultural landscape analysis; this is, as Alexandru Dragomir (2008, p.245) mentioned, not only "a list of endings and beginnings, but also an overlap-ping of such lists". That is to say, an imperfect superposition of the multiple layers of the everyday and of the human reality. We have already stated in, the chapter dedicated to the analysis of metaphor and palimpsest within the cultural landscape that it is fair to discuss again about the layers of human intervention. Alexandru Dragomir (2008, p.198) declared: "how does man create? Myths, legends, environment, philosophy, science, arts of all sorts (poetry, prose, painting, music, sculpture, architecture), techniques, politics, logic. Some respond to needs, others develop these needs, while others (?) come for free. We find our-selves in a human world (in a «closeness») where technique and not nature is the stranger (the object); a world where «nature» is banished (reservations, recreation places, zoos)". Alexandru Dragomir (2003, p.32) therefore thought that these layers of the human creation determine the fact that we are in a world of humanized realities. As a result to all the statements presented, we can distinguish why it is necessary to know the theory behind the concept of cultural landscape although we are still on a relatively poor level. We refer again to one of geographer Michael Jones' statement who considered that cultural landscape is firstly a perspective. It is a perspective as much as it is an idea, and both are based on a series of cultural processes. Alexandru Tănase (1968, p.31) explained that culture "cannot avoid relating to two fundamental realities it is closely linked to: nature and society". This makes nature, culture and society essential components of the everyday, of the Weltanschauung and of the cultural landscape itself.

We consider the concept of cultural landscape as a concept of disintegrating com-plexities or even of complex disintegrations, no matter how we analyse this statement. We take it as a paradox as well as a permissive ambiguity both in theory and in practice. We will complete the already mentioned ideas with a matter we have always treated with attention: the reference to our immediate reality, through theory. In order to explain it, we will bring to attention the question of the specific view of nature and the framework where the cultures of different nations manifested. The lack of a clear description of the aspect the concept of cultural landscape is not valid. Hence the importance of the study and of the knowledge of the concept of cultural landscape which consists in understanding the two terms of the expression.

Throughout the study we quoted Romanian authors such as Constantin Noica, Lucian Blaga, Girgore Georgiu, Alexandru Tănase, Tudor Vianu, Alexandru Dragomir, Augustin Ioan, Alexandru Sandu, Cătălin Sârbu, Adriana Matei, Lucian Drăguţ and many others. The next part of the paper mentions Gheorghe Macarie who gathered in his paper entitled *Geografie literară. Orizonturi spirituale în proza românească* (Literary Geography. Spiritual horizons in Romanian prose), writings of famous Romanian prose writers. This book I first presented to my students within the course of cultural landscape and development held by Mr. Cătălin Sârbu almost four years ago. Although it does not offer a scientific approach on the subject, I considered important to read the writing because of its strong spiritual meaning and I thought all students that have a passion for the sudy of cultural landscape should own a copy. Besides, we have already mentioned that no study of the cultural landscape should estange us from landscape, nature, the general framework where everything came to life.

This fact is frequently forgotten and one tends to discuss about a cultural landscape which has nothing in common with the general framework, except maybe indirectly. In this respect, Gheorghe Macarie (1980, p.XII) stated that "nature purifies and rises, maintains the spectacle of an aesthetical variety that is opposed to the daily «gray», it relaxes and educates. These are old words that people repeats for centuries but which constantly reveal new meanings and demands".

Given these new meanings and demands, we must look at the general framework where cultural landscape originated. And how else could this be done than by becoming familiar with the theory? A theory that no one teaches us how to assimilate, which requires a constant dialogue between the researcher and its subject of research. One must understand the narrative character of the cultural landscape as part of its truthfulness in all dimensions. As Kent Mathewson (1999, p.268) stated, Carl Ortwin Sauer himself was influenced in his work by famous contemporary to him American poets and prose writers such as Lewis Mumford, Robinson Jeffers, Gary Snyder and so on and influenced in his turn, prose writers of the American spirit such as Barry Lopez and not only. Mathewson also noted that a few of Sauer's disciples continued his work through works that influenced many other disciplines (Mathewson, 1999, p.268).

Not knowing the theory can lead to the misunderstanding of the entire concept. We find ourselves today in a *zugzwang* situation towards the theoretical study of the cultural landscape. This name was borrowed from the game of chess where it defines a situation when one of the players is required to make a specific move although he knows that any move he makes will put him in a vulnerable position towards the opponent, thus preferring not to make any move (but it would not be possible). It is the case of such a situation in the study of cultural landscape because we consider it compulsory to make some move for the knowledge of the concept. The analogy is not complete because in our case there is no opponent. Perhaps we are the only opponent and the disadvantage of such a move is determined by the way the so-called game, that is the study of the concept, was treated so far. But a move is necessary. How it will be made is a question of rhythm and probably of choice, while when it will be made is a question of time. The transition, as we discussed earlier, is compulsory. We have two possibilities: being faithful to Carl Ortwin Sauer's concept of "rurality" or passing into the sphere of study of what we named a meta-word and considering the cultural landscape only as part of an understanding and explanation process of the landscape as the object of our conscience / consciousness (Cresswell, 2008, p.5). Either way, the transition implies the assimilation of the theories that gravitate around the concept through a specific shaping of the human and nature's spirit which will develop through continuity. But here we discuss about a shaping that is based on the narrative character and the capacity of expressing a dialogue of the cultural landscape (this was presented in the chapter on chronotopy), while transition based on innovation, (re)invention and tradition must grant the continuity.

In other words, the zugzwang situation is based on a cultural production that "starts at the level of society's behavior and practices" (Chombat de Lauwe, 1982, p.124) and must be put into practice preserving the close relation to nature and the environment where it takes place (either urban or rural), and also of a man-nature relation as balanced as possible. As Chombart de Lauwe (1982, p.129) declared, "all construction is a dialogue, sometimes a happy dialogue, others difficult or tragic", a dialogue between the creator and the beneficiary and also between the man and the environment. The dialogue must take place by keeping in mind the necessity of a transition based on innovation, (re)innovation and tradition. It represents a dialogue not only of the past but also of the present and the future, which thing must not be forgotten by the architects, urbanists and landscapers and applied accordingly. However, the same author, Chombart de Lauwe, stated that this specific dialogue is about to disappear although "it is the expression of a life" (Chombart de Lauwe,1982, p.129).

Thus, the zugzwang situation makes us think about any future move in relation with past moves. If our understanding of the concept of cultural landscape is based on a clear and solid theory which emphasizes its evolution and the conceptual richness earned throughout the years, then the zugzwang position we find ourselves in currently, can easily be changed into an advantage. This advantage would assure us that the transition represents change first of all, and our direct intervention would be linked to a reflection on the theory. Regarding this change, Chombart de Lauwe (1982, p.XIII, in the book's introduction wrote by Dr. Fred Mahler) thought that the most important element is in fact "the process analysis and hence of the contradictions that lead to change". Moreover, he explains two important terms mentioned in his arguments – lived culture and action-culture – believing that they are based on the creator's awareness "related to perception and expression, desires and values, needs and necessities, aspirations and projects which will occur in their relation with the environment and their ability to influence beings, things and institutions" (Chombart de Lauwe 1982, p.XIII and p.145). He also thought that "knowing the practices represents the starting point of cultural works" (Chombart de

Lauwe 1982, p.XIII and p.145). We should now bring to attention our own statement made in the chapter on cronotopy where we also discussed about Josep Muntanola whose opinion was that the study of the space (because this is basically our domain) must focus on the changes that occur, and not on its perpetual character (Muntanola, 2004, p.3). In addition, Immanuel Kant (2011, pp.28-29) stated in 1786 about the Assumed beginning of the history of mankind that "the first needs of the living (4:20) imposed by the change of the lifestyle can now turn against each other. This must be how culture and art occurred, both through entertainment and work (…) The raw progress developed the entire human project". If this is the beginning of culture (at least hypothetically speaking), it means that change was no other that "the contact with a soul matrix. We live and we leave behind the centuries but we are constantly connected to our predecessors, i.e. the feeling that we belong to a land where our community of human souls lived, lasted and which they defended" (Macarie, 1980, p.XII). This is why we insist in this chapter of conclusions as well as along the whole thesis on the importance to assimilate a theory of the concept through change. Because as demonstrated in the chapter on cronotopy, the cultural landscape's capacity to change is due to a system composed of the landscape, the man and the built environment. The study on the cultural landscape should not only refer to establishing certain elements created by man through what we call culture. The study aims at understanding and becoming aware that all these results of culture that we find as part of the environment have their own story which assimilated by the landscape. The only desire to study the cultural landscape must determine us to investigate the theory of such a complex and ambiguous concept. Denise Scott Brown (2010, p.113) believed that "critics who ignore the social and cultural dimension in our thought see only historical sources in our symbolism". Therefore, as architects, as creative and cultural beings, we bring our contribution to the continuity of the cultural dimension through our projects, planning and creations. Chombart de Lauwe (1982) considered the cultural dimensions in discussion the essence of what he calls awakening to culture and makes a parallel between the urban and the rural levels. But we must also point out that "in traditional rural life the peasant was the creator of their own environment depending on the changes they made on each detail of the landscape through their work, creating their own style. The dialogue with nature took place directly and their part of cultural creation also connected him to the flora and the fauna as well as to the models, the systems of expressions and values that belonged to their social group" (Chombart de Lauwe, 1982, p.130).

It is therefore easy to understand why the concept of cultural landscape defined by Carl Ortwin Sauer was assimilated to rural life and the change of paradigm only occurred in the second half of the last century. The dialogue with nature changed nowadays. The architects, the urbanists and the landscapers became the main actors because as Kenneth Frampton (2010, p.65) once pointed out, "we architects have a holistic responsibility to satisfy multiple variables. Some are less measurable than others". This responsibility occurs both in the study of the concept of cultural landscape and also in the relations we and "beneficiaries" too have with the environment.

We are not in a theoretical emptiness when studying the cultural landscape as one might wrongly think. Just as we are not in a theoretical emptiness in terms of the daily life. The explanation to this statement consists in the fact that the environment we live in does not lack problems and needs that require solutions and resolutions. It is in our nature to discover solutions and answers and this continuous seeking is founded on a theoretical part. It is so proven that our experience can only be completed if we associate our external sensory feelings, that is what we call "practice", with the knowledge assimilated during our lives, either through direct experiences or through study. To sustain this idea, we present Gheorghe Macarie's (1980) words he wrote on the last cover of the previously mentioned book: "Romanians prove an original presentation of the feeling for nature compared to the same psychical reality of other nations, the same way that the Romanian space presents different characters and a special view of the mountain, the fields, the waters, a special feeling of the time, which are elements of the man nature relation. Knowing the lands and the people, that are travelling (in real life or with our imagination through literature) in all the regions of the country means discovering and recreating as many spiritual horizons" (Macarie, 1980, p. back cover).

Knowledge is what increases our credibility as future professionals, especially be-cause the study of cultural landscape (not necessarily of the concept, that is not the theoretical part) offers us almost unlimited possibilities. The theory is equally permissive, at least in terms of defining the concept. As shown in the beginning of the paper, there concept enjoys numerous definitions depending on how it is understood. Therefore, the researcher's neutrality is solely an illusion because they must be totally involved in the work they carry out which requires their scientific and spiritual devotion. From our perspective, the study on the cultural landscape and on the concept demands the understanding of the elements and the changes within the landscape based on the researchers' sensitivity, perception, in-terpretation, conscience, memory and identity. Alike any discipline, the involvement must be absolute. In conclusion, according to R. J. Johnston (1986, pp.2-3), in the academic en-vironment, the study is carried out on a horizontal theme as well as on a vertical theme: "on the vertical theme, they investigate interrelationship between people and their environment (nature and created – some argue that the two cannot be separated). On the horizontal theme, the study inter-place inter-relationships, emphasizing the flows of people and of human creations across the earth's surface".

As a result, we observe that the study on the concept of cultural landscape must stay loyal to the primary meaning of landscape as we presented it in this thesis. At present the term of landscape has changed its direction from the initial meaning and we thus began using it in expressions such as mediatic landscape, city landscape, urban landscape, corporate landscape etc. and these are only few of its uses. These facts lead perhaps to a distorted understanding of the concept of cultural landscape because its use often loses its initial meaning. Had theory provided us a more complex and complete explanation on the concept and its evolution, we would have easily made a selection of the study and increased the standards of understanding. We probably mention habits and immaterial culture too often and meet the risk of understanding cultural landscape as whole composed of such elements. It would be an improper approach, but not entirely wrong. We can speak of habits and immaterial culture as long as they are connected to the landscape, its changes, evolution and transition. We will then emphasize a series of customs particular to a community without actually referring to the study on cultural landscape. It could easily be presented by an ethnologist for example and this type of practices were culpable for the concept's distortion and disintegration. We changed it into a concept of complex disintegration and of contradictions because somewhere along the line we understood things completely wrong. Either from the need to establish the boundaries between our work as architects and the work of the geographers, or from a falsely assimilated primary idea on what the concept represented. In order to set to rights the understanding of the concept we took it under the shape it came to life initially and observed its evolution. We must be able to analyse the evolution, criticize it and bring our contribution to its development using our own understanding once we are very familiar with the theory, of course.

10.2 Consequences to knowing the theory of the concept

We pointed out the importance of knowing the theory of the concept of cultural landscape in the previous chapter. We also mentioned that, in our opinion, its use should not stray from the meaning granted initially. What could the consequences of knowing the theory of the concept be? The question may seem awkward because the theoretical study is much too relevant. As Friedrich Wilhelm Joseph von Schelling (quoted in Habermas, 1983, p.121) noticed, "the consequence of leaving aside all theory and being concerned only with practice is the flatness of action as well as of knowledge. The study of a theoretical philosophy makes us aware of the world of ideas, and ideas alone confer energy and moral significance to all actions". Moreover, Juergen Habermas (1983, p.121) stated that "the only type of knowledge which has the power to direct us in our actions is the one which flows from simple interests and makes its way back to ideas". It is precisely the need for knowing and using the theory while its results consist in a validation of ideas born from the understanding of a series of complex and hard to assimilate concepts. The validation could be the acceptance of the ideas generated by theory and also a expression of the interest generated by the study. For no study or action can exist without an interest, be it scientific, spiritual or material, i.e. practical (Habermas, 1983, p.134)[153].

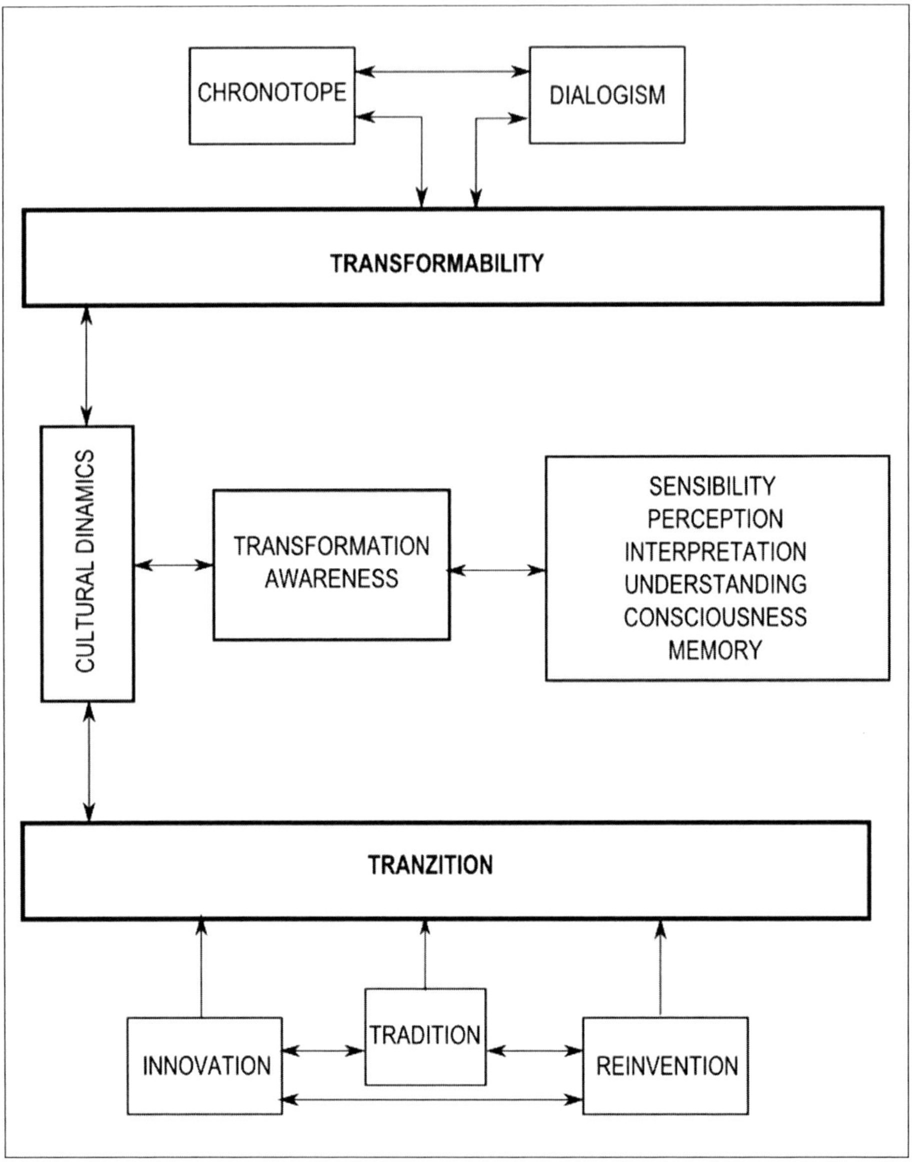

Figure 37. The dynamic of changes and transition in cultural landscape on a conceptual level. Contribution of the author.

We used the concept of knowing (knowledge) in the title although we had in

[153] According to Juergen Habermans "those interests related to the natural history of the species and which we consider to be the starting point of the interests leading knowledge belong at the same time to nature and the cultural separation from nature".

mind the use of the word "understanding", but the knowledge of the concept does not necessarily involve its understanding which is our next concern. The same theoretician Juergen Habermas (1983, p.135) thought that reality is divided in "three categories if possible knowledge" information – which extends our technical ability to possess things, interpretation – which allow the orientation of action within common traditions, and analysis – which frees conscience from its dependence on certain hypostatized powers". Information, interpretation and analysis determine understanding as well as knowledge. The consequences of such knowledge are a series of contributions to the development of the theory and to putting into practice certain efficient methods of practical analysis. As a result, the knowledge of the theory forces us to think, to reflect because it makes visible the effect generated by significant methods of practical analysis.

The knowledge of the theory along with the interest for the study of such a subject and the scientific and spiritual involvement of the researcher translated through its devotion lead to good practice. At the same time, the knowledge of the theory and the evolution of the concept helps us discover new meanings and even definitions. In this thesis, we took the risk that certain conclusions diminish the confusions that the concept brought about during its evolution, most of the times from a will (and need) to define it. It is important to make confusions created by various definitions obvious, as Michael Jones believed. With this effect, one of the obvious consequences of the study on all the theories, definitions and axioms consists precisely in giving efficient definitions of the concept which helps perform our study. The reason being that "a systemic presentation of the different meanings of a word and of their use can lead to better understanding, communication and dialogue through the users' awareness of the differences in understanding the concepts" (Habermas, 1983, p.122)[154]. Besides all this, reading a significant number of papers related to the subject helped us better understand the complexity of the concept and to which extent the authors referred to dedicated themselves to studying the theoretical part of the study of the cultural landscape. As mentioned in the previous subchapter, spiritual involvement becomes compulsory and we bring to attention Lucian Blaga and Constantin Noica's writings. This way the concept of cultural landscape gained several definitions based on the confusions and the contradictions created by its theory. This chapter of conclusions offers a list of definitions we outlined after a close study of the theory. Moreover, we will also present a series of theses, axioms or structure principles which fulfill the role of conclusions which we presented as we advanced with our research.

Thus, based on a number of academic definitions, we established that the con-

---

[154] Juergen Habermas stated: "and when the philosopher watches the immortal ordering, they cannot deny the size of the universe and refrain from recreating it. They create the expression of the proportions they see in the changes occurring in nature just like in the harmonious musical sequences; they are created through mimesis. Through this agreement between the spirit and the order of cosmos, theory thus steps into the practice of life".

cept of cultural landscape is a symbol because of its richness in meanings, and that it is defined by opposite features: it is dynamic although static, a mental construction although a process and a product. Its understanding as a symbol is due to how it is defined and used. We dedicated the whole subchapter 5.6.1 to the understanding of the concept as palimpsest and metaphor which implies an interfusion between the known and the unknown that gave life to confusions and contradictions.

Following, we gave the concept an operative definition: the cultural landscape, under the shape of ideal chronotope (palimpsest) and of social dialogism (metaphor), is the special expression of succession of interactions (transition, change, and temporality) and determined social relations based on cultural processes. Perhaps more than any other chapter the one referring to the chronotope and chronotopy requires the most detailed explanation. We will pay attention to this aspect and bring supplementary explanations by means of listing relevant arguments.

We started with the first definition of the chronotope as pointed by Mikhail Bakhtin (1981, p.84): "we will give the name chronotope (literally, <<time space>>) to the intrinsic connectedness of temporal and special relationships that are artistically expressed in literature". In our definition we linked the fact that it is an ideal chronotope with palimpsest and social dialogism with metaphor. In Bakhtin's opinion time and space are social and historic elements that unify the concept of chronotope. In this respect, we pointed out the existence of the cultural landscape as an (ideal) chronotope of the meaning because of its constantly double aspect in its nature. Therefore, as a concept, the cultural landscape is defined by the nature-culture association as well as the by time-space association and even by the double nature of existence as shown ib Alexandru Dragomir's papers. As for chronotopy, this represents a process through the state of becoming and (trans)forming it injects to the chronotope. Mikhail Bakhtin discussed about the inseparable connection between time and space. We bring an addition to his statement and say that it is closely related to the inseparable connection between nature and culture in a world of multiple cultures. At first, Carl Ortwin Sauer's concept of cultural landscape drew attention from the social character to the physical character of the environment. But this was to be the ideal context for the evolution of the concept; its representation is now social and it would be impossible to understand if we were not familiar with the theory.

In conclusion, we had in mind the possibility of looking at the cultural landscape chronotope as a way to establish the relation between time and space through creation and through their temporal nature as part of our existence. In other words, the chronotope of the human intervention in the cultural landscape is architecture which "has a distinct advantage to affect culture by symbolic means" "through its preoccupation with aesthetics, codes and the spaces of our everyday lives" (Naik and Oldfield, 2010, p.23). We consider that this desire brought to life Carl Ortwin Sauew's initial concept of cultural landscape and John Brinckerhoff Jackson's following studies in the US. That is how the changing nature of the cultural landscape through human intervention defines the dialogic character of the concept given by the landscape man architecture system. The Chronotope represents the reunion of time and space making them inseparable in the cultural landscape context since they are the basis of cultural production just as architecture connects time with the social and cultural structures that determine space (Muntanola, 2004, p.2).

Given the definition and the further additions we can easily create inexistent connections on the theory level throughout the paper. All that has been presented so far gravitates around the terms of transition and change and of a dynamic we are trying to emphasize as the core of the concept of cultural landscape.

Hoping that we have managed to justify our first definition, we will present the second one which we found in this very chapter of conclusions. We therefore consider the concept of cultural landscape a concept of disintegrating complexities or even of complex disintegrations as we have already discussed previously. It is, at the same time, a concept of at least contradictions. For a better understanding of the statement we must remember those arguments we brought for the definition and the understanding of the concept of cultural landscape. We pointed out that the latter was given several methods of interpretation which lead to the following:

- Cultural landscape is a constant construction process which (re)creates itself based on temporariness (according to Tim Ingold).
- Cultural landscape is the object of conscience (according to Tim Cresswell), a heterotopia (according to Michel Foucault) and a mirror of society.
- Cultural landscape is a mental construction (according to Giovanni Maciocco, Iain Robertson, Iain S. Black).
- Cultural landscape is the expression of cultural values and a symbol (according to Backhaus).
- Cultural landscape is the complex expression of a lifestyle as a result of the synergy between cultural and natural elements which creates a specific extremely meaningful struc-ture (according to Cătălin Sârbu).
- Cultural landscape represents a moral order (according to Ezra Park).
- Cultural landscape is a landscape of power (according to Sharon Zukin).
- Cultural landscape is a cultural process (according to Eric Hirsch).
- Cultural landscape is a space of relations and inflexion (according to West-Pavlov).
- Cultural landscape is a perspective more than it is an entity (according

to Michael Jones).

- The idea of cultural landscape is the salvation of the poor idea of landscape (ac-cording to David Jeffrey Blankenship).
- Cultural landscape is the prologue and the epilogue (according to Christopher Tilley).
- Cultural landscape can be viewed as a result of regularly consolidated landscapes (according to Martin Dieterich and Jan van der Straaten, Krut Faegi).
- Cultural landscape is a resource for tourism (according to Ringer and Czepczynski).
- Cultural landscape is a code and its decryption involves understanding cultural and social significance (according to Donald Meinig).
- Cultural landscape comprises matual and cultural elements of daily, common and familiar landscapes (according to Lennon and Mathews).
- Cultural landscape is a mosaic of three elements: natural, physical (resulted from human activity) and models created along the years in landscape (also according to Lennon and Mathews).
- Cultural landscape is an ideal chronotope and a social dialogism (this conclusion is our own contribution to the evolution of the theory).
- Cultural landscape is Weltanschauung, i.e. a worldview. It is also a view of reality (this conclusion is our own contribution to the evolution of the theory).
- Cultural landscape is a meta-word.
- Cultural landscape is a metaphor and a palimpsest with "potentially as many ways of seeing as there are eyes to see" (Jackson, 2003, p.177).

These are only few of the ways to look at the cultural landscape we outlined in our paper.

The above brief listing can only explain why it was referred to as a concept of dis-integrating complexities. It is certain that the more meanings and definitions it gets, the more its disintegration increases. We are talking about a conceptual disintegration which demands a unifying theory.

Do we really need new definitions of the concept? For didactic purposes, the concept requires a less strict definition that encourages our study. All this while the seeking of a definition is performed with full knowledge of the already formulated definitions or at least of the most important of them all that have been adopted by the academic environment. What we can now conclude with is that the concept of cultural landscape offers a too many definitions and means of understanding to reduce it to a single and final definition or to an in situ study of the without knowing its related theory.

The purpose of the present research is not an attempt to convey new methods of re-search or methodologies as some would expect. Our contribution is strictly theoretical just like the title of the paper suggests. We started on an urban level with what urban cultural landscape or local cultural identity signified. This along with our curiosity for the subject inspired our attempt to put at the young students' disposal a useful tool for the theoretical study of the concept. It was extremely difficult to find the writing which would make relevance for our research as most of them were unavailable in libraries or bookshops in our country. This justifies the obstacles a young researcher meets when dealing with such a subject and any helpful information is more than welcome. Most of the papers studied for and mentioned in this thesis must be part of our libraries, be they our personal libraries or the universities'. We consider that in terms of the methods of study, the tools requested for such activities are already available since they are the same as those used in humanistic fields concerned with cultural landscape: architecture, geography, ethnography or anthro-pology. The study of cultural landscape is dependent of Urszula Myga-Piatek's (2008, p.84) humanistic paradigm because it borrows study methods specific to humanistic and social sciences. Urszula Myga-Piatek considers that these are the only ones that can "define cause-result relations occurring between the nature and the culture". The author also calls attention on the idea that "the main aim of modern research is detection of the mechanism of changes of elements of landscape" (Urszula Myga-Piatek's, 2008, p.84). She refers to change and transition again.

Basically, we believe that the main objective of the studies of cultural landscape is becoming familiar with its theory, and afterwards to select the tool provided by each discipline. We must not forget though that these studies are pluridisciplinary and that they are performed by a group made up of researchers, members of the teaching staff and/or students. We point out once again that a good knowledge of the theory will always give us the professional trust required in pluridisciplinary studies. On the same range of ideas, a good practice is always based on a well-built theory. In conclusion, we propose five axioms or theses extracted from the theoretical argumentation in the present paper. These are as follows:

Axiom 1:

The understanding of the concept of cultural landscape can only happen by the understanding of its two composing terms: landscape and culture, on a conceptual level.

Axiom 2:

Firstly we must know the evolution and definitions of the concept of cultural landscape for an efficient use of the concept.

Axiom 3:

The importance of the theoretical study of the concept must be viewed through the disciplines involved in the study and their suitable theoretical tools.

Axiom 4:

In the study of the concept of cultural landscape, the knowledge of concepts of space, place and time supports its understanding as a worldview constantly related to the everyday.

Axiom 5:

The elemental time-space connection induces the existence of the cultural landscape under the form of a chronotope, an essential property we cannot assume unless we have become very familiar with the theory.

The five axioms reveal arguments elaborated in each of the chapters of our book. The arguments of first axiom are presented in subchapters 7.1 and 7.2; the second axiom relates to subchapters 7.3 and 7.4, also with extension for the following two axioms in sub-chapter 7.5. The third axiom is presented in detail in chapter 8, the arguments of the fourth are detailed in subchapters 9.1 and 9.2, and finally, the arguments of the fifth axiom are found in subchapter 9.4.

We will now make a final note as conclusion and reiterate some of the ideas discussed in this very chapter. For start, in our opinion, the researcher cannot be neutral in the theoretical study of the concept of cultural landscape and defining one concept is always related to the other concepts involved in the study. Then, drawing definite conclusions could make the concept of cultural landscape limited and static which is the exact opposite of their real character, originator of confusions. There will never be a single definition of the concept, nor for the terms of landscape and culture. There will never be a fixed theory either. What will be in-deed is a theory based on the awareness of the changes, and numerous definitions formulated in accordance with the already existing definitions and meanings. This is the very scope of this book. We are quite confident that the present book will offer numerous and valuable pieces of information for young architects, urbanists and landscape architects. Finally, the thesis can be looked at as an important starting point in future studies where theory represents a remarkable achievement in a scientific attempt to understand the cultural landscape field of studies. I hope that this book will be of great value, despite perhaps the metaphorical language and sometimes controversial findings, as it is a sincere and sensitive approach on this much disputed subject.

11 Appendix

11.1 About a Central-European Consciousness of Transylvania

Completing this book, we found very adequate to present a study as an applied theory on the concepts of consciousness, specificity and local identity. Therefore, in the following chapter we made a short analysis on the Central-European consciousness of Transylvania, through the theoretical aspects from the related chapters.

11.1.1 Short historical background of the destiny of a geographical suburb

Transylvania is probably the most known trademark of Romania, mostly because of the legends created on top of its historical background. Beside this, Transylvania was al-ways a gateway to/from Europe. In this respect, it is significant for our study to determine how the legacy of the Roman and the German and Austrian Hungarian domination was the source of a Central European consciousness of the land between the Carpathians.

The geographical and early social aspects of Dacia and its people were very well described by Ion Grumeza who stated that "the semi-circle formed by the Carpathian Mountains provided excellent shelter in the elevated plateau of Transylvania, whose name means <<the land beyond the forests>>. With its mild climate, forests abundant with fruits and animals, fertile soil, and ores of salt, gold, and other subterranean riches, Transylvania was populated with groups of wanderers who were forced to share living space and cooperate during their fishing and hunting expeditions. (…) By the middle of the last millennium before our era, they had united into permanent tribes named Daii, Daoi, Dacs, Dacii, Dagae, or simply Dii, names which are all derived from daos, an ancient Greek word for wolf, the animal most worshipped by Neolithic Dacians" (Grumeza, 2009, p.2).

Furthermore, he continued by describing how "the Greeks and Romans were too distant and concerned with their own wars to pay attention to Dacians. To them, the Daci-ans were the barbarian Getians of the western Black Sea shore, people who they misnamed Thracians, after the tribes at the northern border of

Greece" (Grumeza, 2009, p.3). But this didn't stop the eastern expansion of the Roman Empire into the region of the ancient Tran-sylvania, and around the year 100 A.D (in 106 A.D.) the Romans defeated the Dacs and the emperor Traian established the province of Dacia as part of the Roman Empire. After 117 A.D. Hadrian changed the name of the Traian's province from Dacia into Dacia Superior, where the nowadays Transylvania was a key component and formed the very border of the Roman Empire with the barbarians (Bărbulescu, et. al., 1998). Thus, the Romans considered its natural resources and its favorable geographical position as of great interest because "this area was the primary route to both the Near and the Far East, and the place where the barbarians pouring out of Asia had to be restrained. Dacia's natural barrier, the Danube River, made an ideal buffer zone for the western and southern European civilizations" (Grumeza, 2009, p.3).

Thus, we could say that the destiny of Transylvania as geographical suburb and border of an empire was sealed even during those early times. By the year 170, the expansion of the Roman Empire was over and its decline started. This affected the province of Dacia Superior, and by the 4th and 5th centuries, the area of today's Transylvania was no longer part of the Roman Empire. Thus, the civilians who were left behind by the Roman army could no longer maintain the urban fortifications. This was the moment when all the temples and fortifications vanished. The ruralization started as a response action to the urbanization produced by the Romans. Civilians were able to secure only the small rural settlements and not the big fortifications left behind by the Roman army. It was the beginning of a rural state, as a foreshadowing of the Medieval Age (Bărbulescu, et. al., 1998).

Nevertheless, the Romanians known by the name of volohs or valahi, continued to live on the former territory of Dacia, inside the Carpathian arch. Until the 11th century, Transylvania and its people witnessed many wars with barbarians and migratory people that crossed this land or even settled there. The Hungarians settled in the Pannonian planes in the 9th century started their eastern expansion. This was the beginning of a Slavic invasion that was to determine the very future of Transylvania. Important influences are to be seen in Transylvania's toponymy, urban settlements and culture. The invaders imported their culture on top of the Romanians culture, defining by this a new and rather different cultural specificity.

In the 9th and 10th centuries, there are rather few historical documents that testify the existence of the Romanians in Transylvania and even in the Balcans. That information was published for the first time in the 11th century in a chronicle called Gesta Hungarorum. The writer of this document remained the subject of an enigma. It is only known that it was the notary of King Bela. Looking for the truth, Alexandru Madgearu dedicated an entire book to this matter, entitled The Romanians in the Anonymous Gesta Hungarorum. Truth and fiction. The core of this matter is shoun in Alexandru Madgearu's conclusion in elucidating the identity of Gesta Hungarorum's author: "In conclusion, we consider that the present state of the investigation cannot provide a final solution for the identity of the Anonymous Notary. In the first edition of our work we preferred a date around 1150 for the writing of GH, without excluding other possibilities. If we accept the date after Bela II, then the identification with Paulus, bishop of Morisena, would be worthy of consideration. It is nevertheless true that the later dating, after Bela III, is also supported by valid arguments. In this case, the best solution could be the identification proposed by G. Györffy: the prepositu from Buda, active around 1200" (Madgearu, 2005, p.20). Though, as this is not actually the subject of our research, it was worth mentioning Gesta Hungarorum as evidence of the existence of Romanians in the Balkans and Panonnian plains. This fact is of more importance in our quest to determine a Central European consciousness and identity in Transylvania.

In the second part of the 12th century, the Hungarian expansion to the eastern limits of the Carpathians was possible with the help of the Szekely ethnic group (also known as Szeklers or Siculi) of Hungarians and of the German colonists, known by the name of Saxons (very few linguistic studies determined that the first colonists who settled in Transylvania came from the Rine, the Flanders and northern France) (Fabini, 2009, p.13). The geographical location of those two ethnic groups determined different influences in Transylvania. The Germans, known by the name of Saxons, were the ones to defend the southern line of the land, and the Szeklers had to defend the eastern line. The Saxons played a major role in the western influence in Transylvania. They were the ones to build wealthy villages and to create important commercial routes in the area. Thus the urban life in Transylvania started to rise again (Bărbulescu, et. al., 1998).

The long history of the Saxons and Romanians in Transylvania exceeds the terms of this research. The most important aspect for building a Central European consciousness of Transylvania relays in the very first German infusion in the region. In conclusion, in order to complete a short and concise historical framework of the first Saxons in Transylvania, we mention James P. Niessen's presentation of the Germans in Romania which is very con-clusive: "At its height, during the interwar period, Romania's German minority was ten times larger than its current size. The ancestors of some Germans were brought to Transyl-vania by Hungarian kings in the thirteenth century to settle and defend the southern borderland of the province, while others came in the eighteenth century to help revive the economy of southeastern Hungary after its liberation from the Turks. Today perhaps two-thirds of the Germans are Roman Catholic Swabians (Schwaben) residing in the Banat, while less than 20,000 Lutheran Saxons (Sachsen) remain in southern Transylvania. Many Germans fled westward at the end of World War II, and a large portion of those who remained were deported to the Soviet Union. Most deportees returned to Romania in the 1950s but began to emigrate in the 1970s with the collusion of the Romanian and West German governments, which paid a ransom (nominally in remuneration for their education in Romania) for each person. Although the ransom system ended with the fall of communism, German emigration became a flood in 1990. Many abandoned Saxon villages have been occupied by Roma, while their historic churches are preserved by foundations based in Germany or have been purchased by Romanian congregations" (Niessen, 2005, p.742).

11.1.2 About consciousness through the dominant role of the fortified settlements

Right from the beginning of this story, we would like to point out the words of Lászlo Kürti (2001) who stated that "to analyze the contests over Transylvania—the land beyond the forest, for this is what the Latin "Trans-Sylvania" entails—is of paramount importance. This is especially true for both anthropologists and East European specialists, for it provides a way to understand how national movements in their East and Central European settings have privileged the question of national and state borders, and moreover how such national identity movements have affected European and global national movements and politics" (Kürti, 2001, p.3). In this respect, the following arguments are of great importance for our research.

Transylvania is thus the most eastern incubator of Central European identity and culture and this raises the research question about how does a Central European conscious-ness exists in the far eastern borders of Europe? We will not get into nationalist discussions about this contested land between Hungarians and Romanians, but rather construct an intellectual argument on the Central European consciousness of Transylvania and create a framework for a more particular study, the one on the cultural landscape of this remote region.

In this respect, is important to remember that, a Central European affiliation of Transylvania is not (only) political, but more a cultural one. And this determines a geo-graphical nexus. The consciousness is derived from the historical background of the region, but it still resides in its local and regional cultural landscape. As Laszlo Kürti argued "In Europe in general and Transylvania in particular, such discourse has privileged the periphery, described often as a borderland or a border zone, but still as a faraway locale. In fact, Transylvania has been described by many travelers as Europe's Shangri-la. It is viewed not only as a self-contained border culture but is elevated to a special terrain that could best be understood by the phrase suggested by Edwin Ardener, as a remote area" (Kürti, 2001, p.16). The discourse the author describes is the very cultural discourse of the periphery of Europe.

Being at the border is at the same time a positive and negative aspect. Besides feeling marginalized, the border regions have the benefit of neighboring different cultures and therefor earning diversity and specificity. Thus, "as indicated by ethnographers, folklorists, linguists, and writers, Transylvania's regional eccentric culture patterns function as markers or warning posts that something else is different there. Remote regions not only close in and keep out—as we have seen above—they produce their own logic of their own sense of <<being there>>. (...) In fact, border regions are full of (inter)actions, entanglements, and ambivalence, as they represent thresholds of cultures that never end but lead to newer forms of life. As viewed by the contending sides, borders are necessary in fashioning na-tional traditions, for they reveal that dichotomies and hard core facts are rarely justified" (Kürti, 2001, p.20).

For a better way to understand the following regional story and the urban story we found of great help presenting the following quotes: (1) "In its clear account of the part played by the middle class in the development of the modern economic system and modern culture, it makes us realize how much cities of the present have borrowed and are borrow-ing from medieval municipal institutions. It also makes us realize a fact which had been forgotten (…) the importance of the middle class as a revolutionary force". Those are the words from 1946 of Frank D. Halsey the translator of Henri Pirenne's Medieval Cities. Their origins and the revival of trade (Pirenne, 1946, xii). (2) "Another feature of the town, at least of the medieval town, is that its fields, cultivated by its own citizens, could not possibly have produced enough food for their subsistence. It always had to import foodstuffs from areas beyond its direct control. And, because these had to be required of paid for, the citizens necessarily had to produce goods or perform services for outsiders. Every successful town established over the course of time a complex of relationships not only with its surrounding areas, but also with distant places which in turn supplied it with both food and the materials needed for its crafts. This was the hinterland" (Pounds, 2005, p.1).

In this respect, Henri Pirenne stated that there are several stages which define the evolution of the medieval towns and fortifications. First there is the decline of fortifications from the 5th to the 9th century, the revival of military architecture in the 10th and 12th centuries, the apogee of the medieval castles and fortifications in the 13th and 14th centuries, and the transition of the fortified towns in the 14th and 15th centuries.

Transylvania, in this respect, as a remote area and a border zone, had to evolve and to establish a fully operational network of settlements, to connect each other by economical, cultural and religious means. Thus, "every successful town established over the course of time a complex of relationships not only with its surrounding areas, but also with distant places which in turn supplied it with both food and the materials needed for its crafts. This was the hinterland" (Pounds, 2005, p.1). Therefore, the main function of the fortified towns built in Transylvania was to protect the Eastern borders of the empire. They were the outposts, the very first lines of defense. Given the geographical particularities of Transylvania, being itself an enclosed entity inside the Carpathians, the very few passes through the mountains were the ones that had to be defended. And those passes were the Mureş river valley to the West, the Olt river valley to the South and the Prahova river valley to the South-East. Because of that, important towns like Sibiu (Hermannstadt), Braşov (Kronstadt), Sighişoara (Schaessburg), Mediaş (Mediasch) and Bistriţa (Bistritz), among others, flourished under the Saxons and won their urban autonomy along the most important trade routes in the region.

The German medieval urbanism was the base for the fortified churches and towns in Transylvania. The need to defend the borders of the Austrian - Hungarian empire determined the need to fortify the major towns and villages in the area. As a general rule in all Transylvanian medieval urbanism, most fortifications were raised near ecclesiastic entities. Therefore, in medieval towns in Transylvania, besides the fortified church, the fortifications of the town itself (intramuros) came as a necessity, but only by the second generation of local Saxons, thus around the 14th or late 15th century.

Determinants by which the Saxons chose one area or another to settle are not known exactly from historical data. The geography of Transylvania favored the construction of towns and villages on hills and near streams of water. The fertile grounds deter-mined the development of agriculture and animal breeding, and by that time, the Târnave region, for example, gained famous for viticulture. The specific of the land and the knowledge imported by the Saxons led to the construction of the settlements and they are now the proof of the German blood that once flew in the veins of the locals. The birth of a local specificity was then taking place in the very heart of the Carpathians.

The geographical location, combined with the imported cultural identity and specificity of the Saxon settlers plus a little amount of diversity given by the Romanian local cultural identity constitutes the secret recipe for the formation of a distinctive local (even regional) new cultural identity that transcends over Central Europe into its most eastern borders. In this respect, let's remember that "a people has, therefore, a twofold location, an immediate one, based upon their actual territory, and a mediate or vicinal one, growing out of its relations to the countries near them. The first is a question of the land under their feet; the other, of the neighbors about them. The first or natural location embodies the complex of local geographic conditions which furnish the basis for their tribal or national existence. (…) Clearly defined natural locations, where natural barriers of mountains and sea draw the boundaries and guarantee some defined degree of isolation, tend to hold their people in a calm em- location-brace, to guard them against outside interference and infusion of foreign blood, and thus to make them develop the national genius in such direction as the local geographic conditions permit" (Semple, 2010, pp. 92-93). Transylvania, nevertheless, has the perfect attributes that Semple described. Inside the Carpathian Mountains arch, the genius of the locals got born, but, as history showed us, with foreign infusion. Right from those times (12th century), Transylvania was on its way to become a suburb of Central Europe where the German (Saxon) genius, culture and specificity were the foundations for the local identity.

The Saxon settlers from Transylvania were considered by the Hungarian royalty as contributors to the stability of the central power and statehood. This could be obtained by implementing advanced juridical and community forms inside their urban settlements. One important characteristic of the German settlements in Transylvania was the method of split-ting fields by Flemish sessile system (Fabini, 2009, p.12). This means that the agricultural fields were outside the settlement, permitting the latter to develop among one street or around a central square (Fabini, 2009, p.12).

Hermann Fabini describes the colonization of Transylvania as having several stages. The geographical particularities of the land made the settlers to enter Transylvania through the Mureş river valley and settled near nowadays Alba Iulia. In the 12th century, Sibiu was the most important settlement in settlers' route. Then, around 1200, the Teutonic knights settled in Ţara Bârsei, the area around Braşov and had to defend the mountain passes against barbarian invasions from the South and East.

Thus, the entire area between Orăştie to the West and Drăuşeni to the East was now called Siebenbuergen after the seven distinctive Saxon administrative units (Fabini, 2009, p.14). Those units were called scaune in Romanian or Stuehle in German or chairs in Eng-lish. Siebenbuergen was the name for this part of Transylvania and it came from the seven important medieval towns that were Sibiu, Braşov, Mediaş, Sighişoara, Sebeş, Rupea, and Bistriţa

But the Tatars (also known as the Mongols) invasion from 1241 was a test for all recently created administrative units in Transylvania. After defeating the Russians around 1223, in 1241 they invaded Central and Eastern Europe. In Transylvania they destroyed all the northern towns and entered Ţara Bârsei from Moldavia, defeated Braşov and went for Sibiu (Fabini, 2009, p.15). As Michael Angold stated, "a carefully organized sweep into the heart of Europe followed" (Angold, 1997, p.104). After defeating Braşov, they went to Sebeş and Arad, towards Mohi. The two Mongol armies, one that invaded Poland by the North and the one that invaded Transylvania by the South-East met in Hungary with the purpose to invade Central Europe (Angold, 1997, p.104). After the Tatar's sweep, Transylvania was a devastated land where most of the villages and towns had been destroyed and people were starving. Therefore, the Hungarian king Bela The Fourth initiated new colonization campaigns in order to compensate the misery that had been left behind (Fabini, 2009, p.15). In the second part of the 13th century, the colonization started on the valleys of Târnava Mare and Târnava Mică rivers. The two chairs known by the name of Mediaş and Şeica received the title of Libertas Cibiniensis that gave them special legal rights and duties to the royalty and was the starting point for an economic growth in the region (Fabini, 2009, p.15).

There are few historical documents about the existence of a known Terra Medies. By the year 1283 in the area of Mediaş there were eight known villages, all, except one village called Alma, on the left side of Târnava Mare river (Mărculeţ, 2011, p.15). Those were always associated with the name of a small region called Terra Medies or Terra de Medies that was known as the area around Mediaş and the two chairs. By the year 1289 Terra Medies had gained one more village having then a total of 9 (Mărculeţ, 2011, p.16). Terra Medies was considered to be the very heart of Transylvania, in the middle of the old Roman province. In today's Romania, Mediaş is close to the geographical centre of the country: 45°40' – 46°14' northern latitude and 24°20' – 24°35' eastern longitude.

Many historians, geographers and scholars discussed around the origins of the name of this region and its homonym settlement. Some of them traced the name back to the Romans (as Terra Media, the very middle of the province as we stated before), but the analogy with any castrum or army or even hypothetical centre with the name of Media is only plausible and cannot be backed up by any historical proof (Mărculeţ, 2011, pp.19-23). It is more probable that the toponym of Mediaş (Medgyes in Hungarian) has Hungarian roots, as the area of Târnava Mare was first populated by the Szekely in the 12th century and had the role to defend the Empire's south-eastern borders (Mărculeţ, 2011, p.24). This statement is supported by historical documents which testify that all the Toponyms of the settlements in Terra Medies have Slavic roots, and Mediaş is no exception.

As Henri Pirenne stated, "every stable society feels the need of providing its members with centers of assembly, or meeting places. Observance of religious rites, maintenance of markets and political and judicial gatherings necessarily bring about the designation of localities intended for the assembly of those who wish to or who must participate therein. Military needs have a still more positive effect. Populations have to prepare refuges where will be found momentary protection from the enemy in case of invasion. War is as old as humanity and the construction of fortresses almost as old as war." (Pirenne, 1946, p.57).

As a general rule in Central and Eastern Europe, settlements and most of the fortified towns were built in strategic areas that could give a significant advantage in case of attack. The trading routes were the ones that needed stronger defense. Nevertheless, the fortified settlements appeared on natural defense lines like rivers and hills. In the area of Târnava Mare and the region between the Olt and Târnava Mare rivers, there was in place a special defense mechanism blocking any possible enemy attacks from the south. Fortified churches and towns appeared as a specially designed cluster of settlements that had a defensive role. This was most common after the Mongols attacks from 1240s and the Ottoman attacks from late 1430s. This is why many of the fortified establishments from this region are from the late 15th century.

The evolution of every medieval town is, foremost, a proof of its need to defend it-self and its surrounding territory. It determined the specific urban structures of the medie-val settlements, as "security was a major factor in the creation and growth of most towns" (Pounds, 2005, p.25).

But, as Paul Barford argued, "the question of differences in economic development across Europe took on an additional political meaning with the claims of German scholars concerning the civilizing influences of the Ottonian penetration into the region east of the Elbe River. In particular, the adoption of a narrow legalistic definition of a town emphasizing the existence of a charter allowed these scholars to insist that the Slavs came to benefit from the urban civilization very late and only under German influence" (Barford, 2005, p.84). The same is the story about Transylvanian urban settlements, who were based on the German medieval urbanism.

The geopolitical aspects at the beginning of Medieval Ages were definitive in the evolution of the urban settlements in the Eastern European region, and furthermore the very cornerstone in the path of economic growth and social changes and the basic formation of Transylvania's destiny. "The destinies of cities unfolded in the context of the movement of peoples and consequent upheavals which swept away the dominant institutions of the old mode of production, leaving a more flexible situation in which the new could develop" (Southall, 1998, p.95). Furthermore, as in case of any new starting point in history, after the decline of the Roman Empire, the urban settlements of Central Europe continued to exist under the leadership of the Church. This customs were imported by the Saxons and started to get shape in Transylvania as a base for future developments in the urban settlements.

There are several categories and aspects that need to be taken into discussion when a medieval town fortification is analyzed. Among them, the character of the site, the general character of the urban settlement, the structure of the street network, the character of the streets and of the constructed mass and the space covered and the type of architectural components and wholes designed for defense are the most important ones (Gheorghiu, 2000, p.15). The architect Teodor Octavian Gheorghiu managed to fill up the gap of medieval urbanism history in the Romanian literature with a bilingual book called Fortified towns. Urban defense in Medieval Central and Eastern Europe. Thus, based on the five characteristics enumerated before, the author analyzed important medieval towns to deter-mine their own character and to create a knowledge base for the history of urbanism in Transylvania.

When researching about medieval towns, one must take into consideration the fol-lowing aspects: a) "the defense facilities of the site have been without doubt one of the major concerns of all constructors until modern times, especially of the medieval constructors. Hard to reach heights, steep slopes, isolated (in one way or another) terrains, swamps, or islands surrounded by streams of water – such places were usually chosen as locations for most of the medieval architectural settlements" (Gheorghiu, 2000, p.16); b) "the classification of human settlements as open or closed, in other words as fortified or not fortified, is one of the most usual. A whole series of structural determinations, global and detailed ones, are triggered by this element, as well as the manner in which their evolution has been influenced so far. It is quite obvious that, in the Middle Ages especially, the interior urban wall was consecrated as a common defense procedure; hence, this element has been persistently recommended as a condition of urbanism" (Gheorghiu, 2000, p.17); c) "the structure of the street network can be irrelevant for defense, but it can also influence defense. Specifically, the street network can be projected and developed according to the defense objectives" (Gheorghiu, 2000, p.18); d) "the nature of the streets, the structure and capacities of the civil common constructions that constitute the fronts are obviously connected to the previous section. This aspect rightfully belongs to urban morphology. One can identify here formations that permit an independent defense, or others that relied on mutual cooperation for the mutual cooperation for the defense of the area. The secured fronts (…), associated with the labyrinthine trajectories of the street network, proved to be extremely efficient for defense" (Gheorghiu, 2000, p.19); d) "in a large number of medieval towns the internal defense, or the defense of the adjacent area, was ensured, in addition to the surrounding wall, by other constructions: monasteries, inns, courts of the ruler, of nobles or of the local aristocracy, houses of the town committee, inhabited towers, residents of the aristocracy, fortified churches, castles, or even urban areas with additional protection offered by independent fortifications" (Gheorghiu, 2000, pp.19-20). The same architect is the one that concludes that "these elements, in addition to occupying significant sites and leading to results that are specific to the street network, and in general for the local urban morphology, have a two-fold function: one function regards living (religious service, storing etc.), while the other refers to defense"(Gheorghiu, 2000, p.20).

Besides such characteristics related to the urban morphology of the medieval settlements, there were other several important regional and geopolitical factors that led to the predominant urban form in the area. One of them was "the important period of centralization of the Hungarian, Czech and Polish Royalties and of the German Empire, which was favorable to a similar urban policy" (Gheroghiu, 2000, p.41). Secondly, there is "the tendencies towards global organization of the territories, of which large areas originated in expansion, came true thanks to the central power, through the creation of an urban network incorporated into the administration, economy, defense, and state communication" (Gheorghiu, 2000, p.41). In the third place, "the strong infusion of homogenization resulted from the German colonization; this colonization targeted an important proportion of the urban space. Furthermore, the German urban legislation was adopted, either in addition to the local legislation, or as the only legal system valid" (Gheorghiu, 2000, p.41). Also important, as we stated earlier, is the "set of phenomena composing the cultural context, as well as the religious one (Catholic), as both contributed to the creation of a cohesive local mentality. This mentality was preserved thanks to the numerous relations among urban centers, and had a strong impact on the architecture and urbanism in the area" (Gheorghiu, 2000, p.41). Nevertheless there were external influences that changed the overall context of the settlements. The local genius interfered on the imported unified urban factors, at a very detailed level.

As the prototype of the urban settlement, the structure of a town was basic. The core consisted of a fortified ecclesiastic complex, like a church or a monastery, and the settlements of the crafts and trades outside the fortified nucleus, and mostly with a focal point to a main square. The fortification of the outer settlements came later as a need for extra protection of the church, and as a general rule, each tower was used and protected by a craft guild.

Figure 38. Fortified church in Mediaş. The Castle as it was known from the old documents. Copyright: Alexandru Calcatinge.

The local genius we talked about previously in this paper is only a paraphrase of what the engines of urban and economic growth were in medieval times – agriculture, trade, crafts and fellowship. This brings us to Norman Pounds's statement that "the town was a paradoxical institution" (Pounds, 2005, p.9). Indeed it was. After the 10th century and the evolution of feudalism, the same author describes two ways of the feudal power to "reconcile the prescriptive city with the feudal view of society" (Pounds, 2005, p.11). Thus, the first method was to impose a castle on the urban foundation to be able to maintain order and surveillance over the population. The second method was to "absorb the town into the feudal order: to restrict its powers and to make it a source of profit. This was achieved by the grant of a charter of liberties" (Pounds, 2005, p.11). This is what happened with regard to the Libertas Cibiniensis and the towns from the Târnava Mare valley. In the Saxon towns in Transylvania the focal point of the urban settlement was the church. It was the nucleus of the towns that in most of the historical documents was known by the name of castle, even though it wasn't a real one. The Saxon settlements in Transylvania were free entities that nevertheless were committed to the German urban legislation. In this respect, from the perspective of the urban form, we could determine that none of them were spontaneous and that "for both urban and architectural aspects, one can refer state without being mistaken the preexistence of a project, or at least of an anticipated result, the nature of which is now hard to specify" (Gheorghiu, 2000, p.21).

Thus the fortifications of the Saxon towns in Transylvania determined an evolution of the urban areas that was unitary and led to integral fortified towns which continued to improve its defense constantly and where "the additional role in the defense of the castle, fort, fortress, or the fortified church and monastery [was] played by some preexisting elements, or elements that were implanted by the feudal system of the town system inside or outside the town" (Gheoghiu, 2000, p.42). The author gives examples from Transylvania fortified towns like Sibiu, Sebeş, Mediaş and Sighişoara. Thus, "one of the most interesting cases in point is, however, Mediaş, a town that was fortified very late. Until its fortification its efficient defense was ensured by the fortified church downtown, often named in documents as castle. Once fortified, the town started quickly to expand, occupying important areas outside the precinct, and thus entering the category of partially fortified towns" (Gheoghiu, 2000, p.43). The other important town in the river's valley was Sighişoara, where "up to a point its territorial development occurred simultaneously with the develop-ment of a reliable defense, based on strong precinct walls on the plateau. (…) Remarkably, three strongly fortified monasteries participated in the defense of the southern and eastern sides of the fortress; this proves the complexity of the gradual fortification of the urban conglomerate" (Gheorghiu, 2000, p.50).

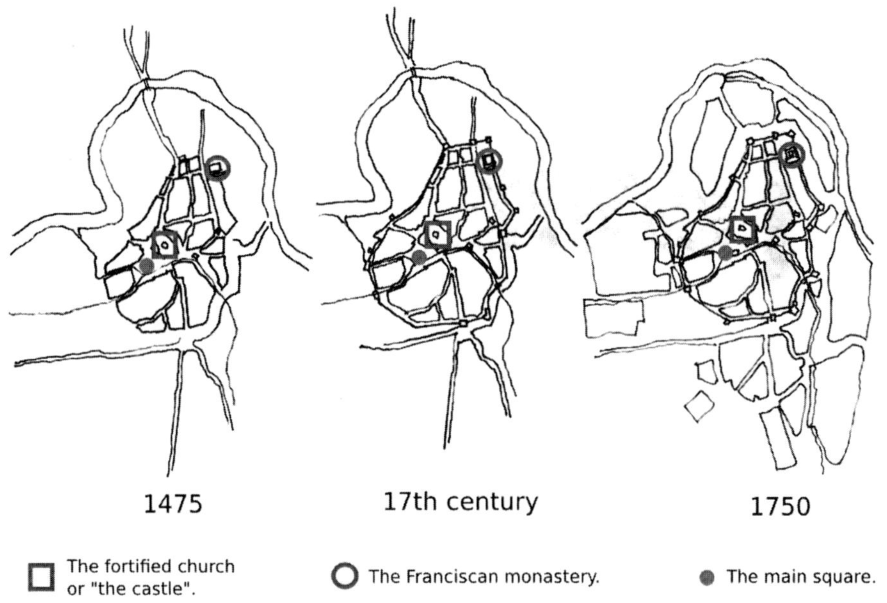

Figure 39. Evolution of the fortification of Mediaş. Copyright: Alexandru Calcatinge.

With regard to the villages and small settlements (or even the extramuros evolution of the fortified towns), the central element in their morphology was the house, that had a special character due to its western genius infusion. Thus, Adriana Matei stated that "the Saxons are building high tethered houses with stone gates, like stronghold walls impregnable from the street. Vertical, aligned, disciplined, the Saxon houses are the expression of their vertical infinity (abyssal). In the same space (the same as landscape) the Romanian houses are alternating disposed, with yard, into an endless rhythm of solid and blind, and the dimension of the volumes is balanced, expressing a sense of measure and of space limitation" (Matei, 2004, pp.59-60). Far from being so philosophical, Paul H. Stahl (2005) created a different description of the Saxon villages, stating that, due to the harsh conditions in the region, the peasants had scattered settlements, usually far away from the travel and commercial routes. Only in the late 18th century the administrative laws determined the peasants to build their houses near the road. But the fortified churches existed even from the 14th century, and thus, they became the central point of the village, with the gathering square nearby. This was like a copy of the already fortified towns that from military reasons had a massive morphological character. Saxon villages were very close to the morphology of the fortified towns, but without the outside walls. Thus "the houses, almost glued one to another, in-line, have as gathering point the fortified churches. The settlement obtains as an ensemble, the character of collective defense, whose functioning and solidarity were based on the strict rules of vicinity - <<Nachbarschaft>>" (Stahl, 2005, p.15).

In this respect, with regard to the urban morphology and urban composition of the settlements in Transylvania, some important aspects can be sketched, as Teodor Octavian Gheorghiu (2000) argued. Those towns had a distinctive role and were different from the ones in the extra-Carpathian regions. In the intra-Carpathian settlements, the urban area had a dominant construction that usually was the fortified church or the monastery, and the rest of the living spaces had a very rural aspect, with few distinctions in case of cleric houses. The difference between towns and villages, once more, was the existence of a specially designed street network, and of course, the existence of the fortifications. But those settlements were outposts on the border of the Empire. Mediaş is an example of a town where the dominant was the fortified church, or the castle, and a second one, the existence of a monastery. One other aspect in the urban composition, besides the dominant castle, was the geographical characteristics of the site, as the town was constructed on a hill near the river Târnava Mare. Those two were used as natural barriers in case of attacks.

At the urban morphology level, we can distinguish the concentration of the street network to a focal point that in most of the cases is the fortified church. In Mediaş, many streets lead to the square in front of the castle, and the fortified church is the major head of perspective on all the streets. Furthermore the streets have a zigzag front for the purpose of defense. This gives a special conformation of the streets; the intersections or inflections are narrow in order to give a better way to defend those points. Entrances to the central square are also narrow and the fronts of the streets are continuous and strongly secured. The fortified church is a spontaneous model, as most of the churches in Transylvania, but the Franciscan monastery in Mediaş is based on pre-established pattern.

Figure 40. View over the old town of Mediaş. Copyright: Alexandru Calcatinge.

The evolution of urban plots and urban morphology in Mediaş is not very clearly documented in historical maps, at least not for the early years, as no relevant documents of this type could be found. According to Avram and Giura (1985, pp.44-57), first houses are believed to have been built around 13th century nearby the present fortified church, at the north, on the highest point of the hill and were occupied by Romanians. The first Saxon colonists are considered to have settled at the south of the present fortified church. At the end of the 13th century, other settlements of the Saxons appeared on the north side together with the local ones and then on the west of the present fortified church. At the beginning of the 14th century and the end of the 13th century the plots of the Saxons and Romanians appeared at the north side of the present church, at the west and south where the Saxons settled, and at the east. Mostly in the first part of the 14th century, on the highest part of the hill very close to the present church, the plots appeared and they closed the entire area. That same time a first church, different from the one that exists now was built on the same spot. At the end of the 14th century, the east and south-east sides of the town are completed and the plots closed the entire area around the church, while the second round of fortified walls that surrounded the church were built. Around 1444, the Franciscan monks have settled in the north-eastern side of the location of the Franciscan monastery. Parallel with the plots, the fortification of the entire area (1490) started and was finished around 1534.

Figure 41. A winter glimpse of a beautiful Transylvanian cultural landscape. Copyright: Alexandru Calcatinge.

As a conclusion on the matter of urban composition and morphology of medieval towns, we once again appeal to the writings of Teodor Octavian Gheorghiu, who stated at the end of his research, that "I have been trying to disprove, once and for all, a belief that has become a certainty with many experts, that Romanian towns are, more than others the result of random, disorganized, unconscious human activity" (Gheorghiu, 2000, p.136).

In spite of history's continuous transformations of the settlements inside the Carpathian Mountains arch, a part of our specificity is the heritage of the fortified medieval towns and villages as a component of a Transylvanian identity that has Central European roots. Furthermore, some special characteristics with regard to local urban and architectural specificity can be added, as the settlements of Transylvania are, besides many other villages and towns around the country, icons and markers of the cultural diversity and spiritual and moral integrity of the people.

But, as Romanian philosophers like Emil Cioran, Lucian Blaga and Constantin Noica argued in many of their writings, Romania has an "unsatisfied consciousness" (Noica, 2010, p.34) and this is perhaps the reason why there is an almost endless search for identity and rebirth of specificity. This discourse is of a greater importance in the contemporary series of events as the need to (re)define our values and construct a healthy and powerful consciousness is essential for the [almost] lost battle of multiculturalism. And what way to end this short intellectual exposition could be better than the words of Constantin Noica, who stated that: "unsatisfied, that is how we [Romanians] basically feel. (...) Unsatisfied with what we know we could be and we are not yet" (Noica, 2010, p.34).

Figure 42. Classical Romanian mountain cultural landscape. In this case, a scenery from the Apuseni Mountains. Copyright: Alexandru Calcatinge

Figure 43. The beauty of a Romanian cultural landscape. Copyright: Alexandru Calcatinge.

Figure 44. View over the Mureş river valley, near Alba Iulia, Romania. Copyright: Alexandru Calcatinge.

Nevertheless, judging by the images above (and from all the images in this book), we believe that there should be only satisfaction, at least with regard to the cultural landscape that we have. And, on top of all this, the history and the local and imported genius that were the base of the constructed genius loci in Transylvania, are unique in Europe (at least).

12 List of Figures

13 Bibliography

- Aalen, F. H., 2004. The study and management of Europe's landscapes. In: Panagiotis Doukellis and Lina Mendoni, *Perception and Evaluation of Cultural Landscapes*. Athens: National Hellenic Research Foundation.
- Aaltonen, M., 2007. Chronotope Space: managing the complex trade-offs between the properties of the strategic landscape and the time frame being considered. In: Foresight, Vol. 9, No. 4, pp.58-62.
- Adorno, T. W., 2001. *Kant's Critique of Pure Reason*. Stanford, California: Stanford University Press.
- Adorno, T. W., 2007. *Minima moralia*. Bucureşti: ART.
- Afloroaei, Şt., 2008. *Metafizica noastră de toate zilele*. Bucureşti: Humanitas.
- Alexander, J. C., 2001. Dezbateri analitice: cum înţelegem anatomia relativă a culturii. In: Alexander, J. C., 2001. *Cultură şi Societate. Dezbateri contemporane*. Iaşi: Institutul European. pp. 5-29.
- Angold, M., 1997. The Mongol-Tatar invasions of the thirteenth century and their impact on the West. In: A. Mackay, D. Ditchburn, (ed.) 1997. *Atlas of Medieval Europe*. London: Routledge.
- Arendt, H., 2007. *Condiţia umană*. Cluj-Napoca: Ideea Design.
- Aristotel, 1959. *Politics*. Cambridge: Harvard University Press.
- Aristotel, 1966. *Fizica*. Bucureşti: Editura Ştiinţifică.
- Aristotel, 2008. *Physics*. Oxford: Oxford University Press.
- Aristotel, 2008. Politics, Book III, Chapter 3. In: Meagher, S. M., (ed.) *Philosophy and the City*. Albany: State University of New York Press, pp. 33-39.
- Armitage, J., 2001. *Virilio Live. Selected Interviews*. Londra: Sage Publishing.
- Avram, Al., Giura, L., 1985. *Mic îndrumar turistic. Mediaş*. Bucureşti: Editura Sport-Turism.
- Backhaus, G., Murungi, J., 2009. Introduction I: The problematic of grounding the significance of symbolic landscapes. In: Backhaus, G., (ed.), 2009. *Symbolic Landscapes*. Springer. pp. 3-33.
- Baker, R. H. A., 1996. The identifying of spaces and places. In: Dominique Vanneste, 1996. *Space and Place. Mirrors of social and*

cultural identities? Studies in historical geography. Acta Geographica Lovaniensa. Vol. 35. pp.1-3.

- Bakhtin, M. M., 1981. *Dialogic Imagination*. Translated by M. Holquist and C. Emerson. Austin: University of Texas Press.
- Bakhtin, M. M., 1993. *Toward a Philosophy of the Act*. Translated by V. Liapunov. Austin: University of Texas Press.
- Ballantyne, A., 2002. *Architecture. A very short introduction*. Oxford: Oxford University Press.
- Ballantyne, A., 2004. *Architectures. Modernism and after*. Oxford: Blackwell Publishers, 2004.
- Ballantyne, A., 2007. *Deleuze and Guattari for Architects*. London: Routledge.
- Banham, R., 1960. *Theory and Design in the first Machine Age*. London: Architectural Press.
- Barford, P. M., 2005. Silent centuries: the society and economy of the Northwestern Slavs. In: F. Curta, (ed.) 2005. *East Central & Eastern Europe in the Early Middle Ages*. Michigan: The University of Michigan Press, pp.60-103.
- Bartetzky, A., Schalenberg, M., 2009. Shapes of Happiness. In: Bartetzky, A., Schalenberg, M., (ed.) 2009. *Urban Planning and the Pursuit of Happiness*. Berlin: Jovis Verlag GmbH, pp. 7-17.
- Baudelaire, Ch., 1863. The Painter of Modern Life. [online] Appeared at Phaidon Press. Translated by Johnathan Mayne. Accesible at the address: http://www.columbia.edu/itc/architecture/.../pdfs/.../Baudelaire.pdf [accessed in 14.09.2010].
- Bărbulescu, M. et. al., 1998. *Istoria României*. Bucureşti: Editura Enciclopedică.
- Bell, D., 1978. *The Cultural Contradictions of Capitalism*. New York: Basic Books Inc. Publishers.
- Bell, D., 2001. Modernism, postmodernism şi declinul ordinii morale. In: Alexander, J. C., and Seidman, S., 2001. *Cultură şi societate. Dezbateri contemporane*. Iaşi: Institutul European.
- Bell, S., 2004. *Elements of visual design in the landscape*. New York: Spon Press.
- Bemong, N., Borghart, P., 2010. State of the art. Bakhtin's theory of the literary chronotope: reflections, applications, perspectives. In: Bemong, N., et. al., 2010. *Bakhtin's theory of the literary chronotope: reflections, applications, perspectives*. Gent: Academia Press, pp. 3-17.

- Benjamin, W., 1999. *The Arcades Project*. London:The Belknap Press of Harvard University Press.
- Benjamin, W., 2007. *Illuminations*. New York: Schocken Books.
- Bennett, T., 2001. *Differing diversities. Cultural policy and cultural diversity*. Strasbourg: Council of Europe Publishing.
- Bird, G., 2006. *The revolutionary Kant: a commentary on the Critique of pure reason*. Chicago: Open Court Publishing Company.
- Birks, H. J. B., 1999. Introduction. In: Birks, H. H., (ed.) et. al., 1999. *The Cultural Landscape. Past, Present and Future*. New York: Cambridge Press, pp.7-11.
- Black, S. I., 2003. (Re)reading architectural landscapes, In: Robertson, I., Richards, P., 2003. *Studying cultural landscapes*. Londra: Arnold Publishers. pp. 19-46.
- Blaga, L., 2011. *Trilogia culturii*. Bucureşti: Editura Humanitas.
- Blankenship, J. D., 2011. *Reading Landscape: Mid-century modernism and the landscape idea*. Amherst UMass. Open Access Disertations. Paper 324.
- Boboc, Al., 2006. *Filosofi contemporani. Fenomenologie, hermeneutică şi ontologie*. Cluj-Napoca: Editura Grinta.
- Boia, L., 2001. *History and Myth in Romanian Consciousness*. Budapest: Central European University Press.
- Bollnow, O. F., 2011. *Human Space*. Londra: Hyphen Press.
- Bondi, V., ed., 1995. *American decades:1970-1979*. Detroit: Gale Research Inc.
- Bonnemaison, J., 2005. *Culture and Space. Conceiving a new Cultural Geography*. Translated from French by Josee Penot-Demetry. London: I.B. Tauris & Co. Ltd.
- Brace, C., 2003. Landscape and Identity. In: Robertson, I., Richards, P., 2003. *Studying cultural landscapes*. Londra: Arnold Publishers. pp. 121-138.
- Breton, A., 1924. *Manifesto of Surrealism*. [online] Accessible at the address: http://www.tcf.ua.edu/Classes/Jbutler/T340/SurManifesto/ManifestoOfSurrealism.htm [accessed in 20.08.2010].
- Breton, A., 1948. *On the road to San Romano*. [online] Accessible at the address: http://www.surrealists.co.uk/bretongallery.php. [accessed in 21.08.2010].
- Bridge, G., Watson, S., 2000. City Imaginaries. In: Bridge, G. and

Watson, S., (ed.), *A Companion to the City*. Oxford: Blackwell Publishers Ltd. pp. 7-18.

- Brown, D. S., 2010. *Architecture words no.4: Having words*. Londra: AA Publications.
- Brown, R., Renshaw, P., 2006. Positioning students as actors and authors: a chronotopic analysis of collaborative learning activities. In: *Mind, Culture and Activity*, vol. 13. no. 3. pp. 247-259.
- Buttimer, A., 1993. *Geography and the Human Spirit*. Baltimore: The Johns Hopkins University Press.
- Calcatinge, A., 2011. *Visions of the Real. An architect's approach on cultural landscape studies*. Vienna: LIT Verlag.
- Calvino, I., 1974. *Invisible cities*. Orlando: Harcourt Brace & Company.
- Carrier, J., 2003. Mind, Gaze and Engagement: Understanding the Environment, Journal of Material Culture, Volume 8 (5), pp. 5-23.
- Casey, S., E., 2010. Aesthetic Experience. In: Sepp, H., R., Embree, L., 2010. *Handbook of phenomenological aesthetics*. Heidelberg: Springer, pp. 1-9.
- Chambart de Lauwe, P. H., 1982. *Cultura şi puterea*. Bucureşti: Editura Politică.
- Choay, F., 2011. *Pentru o antropologie a spaţiului*. Translated from French by Kazmer Kovacs. Bucureşti: Urbanismul.
- Chtcheglov, I., 1953. *Formulary for a New Urbanism*. [online] Accessible at the address: http://www.bopsecrets.org/SI/Chtcheglov.htm. [accessed in 24.08.2010].
- Clark, T., 2002. Martin Heidegger. London: Routledge.
- CoE, Tratat 2000/ETS No. 176/CoE din 20.10.2000. *Convenţia europeană a peisajului*, [online] Accesible at the address: www.coe.int/t/dg4/cultureheritage/heritage/landscape/versionsconvention/romanian.pdf.
- Compagnon, A., 1998. *Cele cinci paradoxuri ale modernitatii*. Cluj-Napoca: Editura Echinocţiu.
- Cook, I., et. al., 2005. Positionality/Situated Knowledge. In: David Atkinson et. al., 2005. *Cultural Geography. A critical dictionary of key concepts*. London: I.B. Tauris & Co Ltd.
- Cosgrove, D., 1998. *Social formation and symbolic landscape*. Wisconsin: The University of Wisconsin Press.
- Cosgrove, D., 2008. *Geography and Vision. Seeing, Imagining and*

Representing the World. London: I.B. Tauris & Co Ltd.

- Coverley, M., 2010. *Psychogeography*, Herts: Pocket Essentials.
- Crang, M., 2003. Cultural Regions and their uses: the interpretation of landscape and identity. In: 1st Conaculta Conference on Regional Cultures, 2003, Quereterao, Mexico.
- Crary, J., 2002. Spectacle, attention, counter-memory. In: Tom McDonough, 2002. *Guy Debord and the Situationist International: texts and documents*. Cambridge: MIT Press.
- Cresswell, T., 2008. Landscape and the obliteration of Practice (2003). In: Nuala C. Johnson, *Culture and Society. Critical Essays in human geography*. Aldershot: Ashgate Publishing Ltd.
- Crouch, D., 1998. The street in the making of popular geographical knowledge. In: Fyfe, R. N., 1998. *Images of the Street. Planning, identity and control in public space*. London: Routledge.
- Czepczynski, M., 2008. *Cultural landscapes of post-socialist cities: representation of powers and needs*. Hampshire: Ashgate Publishing Ltd.
- D'Angelo, M., 2000. *Cultural policies in Europe: local issues*. Strasbourg: Council of Europe Publishing.
- Davis, T., 2010. Third Spaces or Heterotopias?, Sociology, volume 44 (4), pp. 661-677.
- De Botton, A., 2010. *Despre farmecul lucrurilor plictisitoare*. Bucureşti: Editura Humanitas.
- De Meyer, et. al., 1999. *The Urban Condition: Space, Community and Self in the Contemporary Metropolis*. Rotterdam: 010 Publishers.
- Debord, G., 1955. *Introduction to a critique of urban geography*. [online] In: Les Levres Nues No.6 (1955). Accesible at the address: http://www.monoculartimes.co.uk/city-tours/psychogeography/urbangeography.shtml [accessed in 18.08.2010].
- Debord, G., 2002. *The Society of the Spectacle*. Canberra: Hobgoblin Press.
- Denevan, W. M., Mathewson, K., 2009. *Carl Sauer on Culture and Landscape. Readings and Commentaries*. Louisiana State University Press.
- Dewey, J., 2008. Philosophy and civilization. In: Meagher, S. M., ed. *Philosophy and the City*. Albany: State University of New York Press. pp. 110-114.
- Diaconu, M., 2011. Matter, Movement, Memory. Footnotes to an Ur-

ban Tactile Design. In: Diaconu, M., et. al., 2011. *Senses and the City. An interdiciplinary approach to urban sensescapes*. Vienna: Lit Verlag, pp. 13-33.

- Dieterich, M., van der Straaten, J., 2004. *Cultural landscapes and land use*. New York: Kluwer Academic Publishing.
- Dilthey, W., 2001. Studiile Umaniste. In: Alexander, J. C., 2001. *Cultură şi Societate. Dezbateri contemporane*. Iaşi: Institutul European. pp. 33-40.
- Dodge, A. W., 2007. Black Rock. *A zuni cultural landscape and the meaning of place*. Jackson: University Press of Mississippi.
- Doicescu, O., 1983. *Despre arhitectură. Scrieri, cuvintări*. Bucureşti: Editura Tehnică.
- Dragomir, Al., 2008. *Seminţe*. Bucureşti: Editura Humanitas.
- Drăguţ, L., 2000. *Geografia peisajului*. Cluj-Napoca: Presa Universitară Clujeană.
- Edwards, J., 2009. *Language and Identity*. Cambridge: Cambridge University Press.
- Einstein, A., 2010. *Teoria relativităţii pe inţelesul tuturor*. Bucureşti: Humanitas.
- Eliade, M., 2000. *Sacrul şi profanul*. Bucureşti: Humanitas.
- Elliot, B., 2010. *Constructing community: configurations of the social in contemporary philosophy and urbanism*. London: Lexington Books.
- Ellsworth, E., 2005. *Places of learning: media, architecture, pedagogy*. Oxon: Routledge.
- Fabini, H., 2009. *Universul cetăţilor bisericeşti din Transilvania*. Sibiu: Editura Monumenta.
- Faegri, K., 1999. Preface. In: Birks, H. H., (ed.) et. al., 1999. *The Cultural Landscape. Past, Present and Future*. New York: Cambridge Press, pp. 1-7.
- Folch-Serra, M., 1990. Place, voice, space: Mikhail Bakhtin's dialogical landscapes. In: Environment and Planning D: Society and Space, 1990, vol. 8, pp. 255-274.
- Foti, M. V., 2003. *Vision's invisibles. Philosophical explorations*. Albany: State University of New York Press.
- Foucault, M., 1977. The confession of the flesh. In: Foucault, M., 1988. *Power/Knowledge: selected interviews and other writings*, 1972-1977. Edited by Colin Gordon. Random House USA Inc., pp.194-228.
- Foucault, M., 1996. *Cuvintele si lucrurile*, Bucuresti: Editura Univers.

- Foucault, M., 2001. Altfel de spaţii. In: Michel Foucault, *Theatrum Philosophicum*. Cluj-Napoca: Casa Cărţii de Ştiinţă.
- Frampton, K., 2007. *Modern Architecture*. A Critical History. London: Thames & Hudson.
- Frisby, D., 2007. *Cityscapes of Modernity*. Cambridge: Polity Press.
- Ganser, A. Pühringer, J. And Rheindorf, M., 2006. *Bakhtin's chronotope on the Road: Space, time and Place in Road Movies since the 1970s*. Facta Universitatis. Linguistics and Literature, 4 (1), pp.1-17.
- Gelven, M., 1989. *A commentary on Heidegger's Being and Time*. Rev. ed. Illinois: Northern Illinois University Press.
- Georgiu, G., 1997. *Naţiune cultură identitate*. Bucureşti: Editura Diogene.
- Gheorghiu, T. O., 2000. *Cetăţile oraşelor. Apărarea urbană In centrul şi estul Europei In Evul Mediu. Fortified Towns. Urban defense in Medieval Central and Eastern Europe*. Translated from Romanian by Andreea Deciu. Bucureşti: Editura Simetria.
- Giedion, S., 2002. *Space, Time and Architecture. The growth of a New Tradition*. 5th ed., 14th printing. Cambridge: Harvard University Press.
- Giraldi, L., 2005. Cultural Planning: an integrated approach to regeneration. In: INHERIT Symposium, 2005 Goteborg.
- Glassie, H., 2000. *Vernacular Architecture*. Bloomington: Indiana University Press.
- Glazer, M., 2010. *Structuralism*. [online] Accessible at the address: http://www.utpa.edu/faculty/mglazer/Theory/structuralism.htm [accessed in 08.03.2010].
- Gomez, B., Jones, J. P., 2010. Introduction. In: Gomez, B., Jones, J. P., 2010. *Research methods in geography. A critical introduction*. Oxford: Wiley Blackwell. pp. 1-7.
- Gordon, E., et. al., 2008. Placeworlds: Using Virtual Worlds to Foster Civic Engagement, Space and Culture, volume 11, pp. 204-221.
- Groth, P., 1997. Frameworks for Cultural Landscape Study. In: Groth, P., şi Brassi, T. W., (ed.), 1997. *Understanding Ordinary Landscapes*. New Haven: Yale University Press. pp. 1-21.
- Grumeza, I., 2009. *Dacia. Land of Transylvania, cornerstone of Ancient Eastern Europe*. Lanham: Hamilton Books.
- Guga, Şt., 2010. *„Urbanism situaţionist" sau situaţionism urban*. [online] In: Blogul de urbana (24.07.2010). Accessible at the address: http://socasis.ubbcluj.ro/urbanblog/?author=8. [accessed in 2.09.2010].

- Habermas, J., 1983. *Cunoaştere şi comunicare.* Bucureşti: Editura Politică.
- Hakkarainen, K., et. al., 2010. Epistemic mediation, chronotope and expansive knowledge practices. Hong Kong: 9th Conference of Computer-Supported collaborative learning, 4-8 july 2011.
- Halfacree, K., 2003. Landscapes of rurality: rural others/other rurals. In: Robertson, I., Richards, P., 2003. *Studying cultural landscapes.* Londra: Arnold Publishers. pp. 141-163.
- Halle, J. A., 2011. *Construind idei. O introducere In teoria arhitecturii.* Bucureşti: Paideia.
- Hammond, M., Howarth, J., Keat, R., 1991. *Understanding phenomenology.* Oxford: Blackwell.
- Harrison, R., 2009. Excavating Second Life: Cyber-Archaeologies, Heritage and Virtual Communities, Journal of Material Culture, Volume 14, pp. 75-106.
- Healy, M., 2011. *Study Guide. The cultural landscape.* New Jersey: Prentice Hall.
- Hegel, G. W. F., 2006. *Filosofia ca sistem şi sistemul ca istorie.* Translated from German by Alexandru Boboc. Bucureşti: Paideia.
- Heidegger, M., 1971. Building Dwelling Thinking. In: Heidegger, M. *Poetry, Language, Thought.* London: Harper & Row, pp. 143-161
- Heidegger, M., 1977. The Age of the World Picture. In: Lovitt, W., (ed.), 1977. *The Question Concerning Technology and Other Essays.* New York: Harper Torchbooks, pp. 115-154.
- Heidegger, M., 2003. *On the Way to Language.* New South Wales: HarperCollins Publishers Ltd.
- Heidegger, M., 2006. *Probleme fundamentale ale fenomenologiei.* Bucureşti: Humanitas.
- Helleland, B., 2009. Place names as a means of landscape identity. In: Jordan, P., et. al., *Geographical Names as a Part of the Cultural Heritage.* Vienna: Wiener Schriften zur Geographie und Kartographie, Band 18, Institut fur Geographie und Regionalforschung der Universitaet Wien, pp. 25 – 33.
- Hernik, J., 2010. Management of knowledge within the international research project 'Cultural Landscapes'. In: Bloemers, T., (ed.) et. al., 2010. *The Cultural Landscape and Heritage Paradox. Protection and Development of the Dutch Archaeological-Historical Landscape and its European dimension.* Amsterdam: Amsterdam University Press, pp.565-577.

- Heynen, H., 1999. *Architecture and modernity: a critique*. Cambridge: MIT Press.
- Hezekiah, G., A., 2010. *Phenomenology's material presence: video, vision and experience*. Bristol: Intellect.
- Hirsch, E., 1995. *Anthropology of Landscape*. Oxford: Clarendon Press.
- Hollander, J., et. al., 2009. Commentary: Virtual Planning: Second Life and the Online Studio, Journal of Planning Education and Research, pp. 108-113.
- Holquist, M., 2002. *Dialogism. Bakhtin and his world*. Second Edition. London: Routledge.
- Hubbard, P., 2006. *City. Key ideas in geography*. New York: Routledge.
- Hubbard, P., et. al., 2005. *Thinking Geographicaly. Space, Theory and contemporary Human Geography*. Londra: Continuum Books.
- Inglis, D., 2010. Civilizations or Globalization(s)?: Intellectual Rapprochements and Historical World-Visions, European Journal of Social Theory, Volume 13, No. 1, pp. 135-152.
- Ingold, T., 1993. The Temporality of the Landscape. In: World Archaeology, Vol. 25, No.2, 1993.
- Ingold, T., 2007. *Lines: a brief history*. Oxon: Routledge.
- Inwood, M., 1999. *A Heidegger Dictionary*. Oxford: Blackwell Publishers.
- Ioan, A., 2002. Arhitectura interbelica si chestiunea identitatii colective. In: Caietele Echinox 3/2002, Fundatia Culturala Echinox, pp. 80-91.
- Ioan, A., 2006. *Cyberspace ca heterotopie*. [online] In Atelier Liternet published in 07.07.2006. Accessible at the address: http://atelier.liternet.ro/articol/2260/Augustin-Ioan/Cyberspace-ca-heterotopie.html [accessed in 01.07.2011].
- Ioan, A., Mihali, C., 2009. *Dublu tratat de urbanologie*. Cluj-Napoca: Editura Ideea.
- Jackson, J. B., 1984. *Discovering the Vernacular Landscape*. New Haven and London: Yale University Press.
- Jackson, J. B., 1994. *A Sense of Place, a Sense of Time*. New Haven and London: Yale University Press.
- Jackson, P., 1998. Domesticating the street: the contested spaces of the high street and the mall. In: Fyfe, R. N., 1998. *Images of the Street.*

Planning, identity and control in public space. Londra: Routledge.

- Jackson, P., 2003. *Maps of Meaning. An introduction to cultural geography*. Londra: Routledge.
- Jaguaribe, B., 2005. The Shock of the Real: Realist Aesthetics in the Media and the Urban Experience, Space and Culture, Volume 8, pp.66-82.
- James, P. E., Martin, G., 1981. *All possible worlds: a history of geographical ideas*. New York: John Whiley and Sons.
- Jameson, F., 1998. *The Cultural Turn. Selected writings on the Postmodern*. London: Verso.
- Johnston, R. J., 1986. *Philosophy and human geography: an introduction to contemporary approaches*. Baltimore: Edward Arnold Publishers.
- Johson, P., 2006. Unravelling Foucault's 'different spaces', History of the Human Sciences, Volume 9 (4), pp. 75-90.
- Jones, M., 2003. The concept of Cultural Landscape: discourse and narratives. In: Palang, H., Fry, G., 2003. *Landscape Interfaces. Cultural Heritage in Changing Landscapes*. Dordrecht: Klenver Academic Publishers, pp. 21-51.
- Jordan, P., 2009. Place names as ingredients of space-related identity. In: Jordan, P., et. al., *Geographical Names as a Part of the Cultural Heritage*. Vienna: Wiener Schriften zur Geographie und Kartographie, Band 18, Institut fur Geographie und Regionalforschung der Universitaet Wien, pp. 33-39.
- Jordan, P., 2011. The endonym – name from within a social group. In: Jordan, P. et. al., *Trends in Exonym Use*. Hamburg: Verlag Dr. Kovac, pp. 9-20.
- Kant, I., 1998. *Critica raţiunii pure*. Bucureşti: Editura IRI.
- Kant, I., 2000. *Critique of Pure Reason*. Cambridge: Cambridge University Press.
- Kant, I., 2004. *Prolegomena to Any Future Metaphysics that will be able to come forward as Science*. Cambridge: Cambridge University Press.
- Kant, I., 2011. *Inceputul şi sfârşitul*. Translation by Martin Zick. Bucureşti: Editura ALL.
- Kaplan, F., 2009. *1959: The year that everything changed*. New Jersey: Wiley and Sons Inc.
- Kriegel, U., 2007. Philosophical theories of consciousness: contemporary Western perspectives. In: P. D. Zelazo, M. Moscovitch and E.

Thompson, ed. 2007. *The Cambridge handbook of consciousness*. New York: Cambridge University Press, pp.35-67.

- Krinke, R., 2005. Introduction. Contemplation, landscapes of contemplation, contemporary. In: Krinke, R., *Contemporary landscapes of contemplation*. Oxon: Routledge, pp. 1-13.
- Kunzmann, K., R., 2004. Culture, creativity and spatial planning. The Town Planning Review, 75 (4), Liverpool University Press, pp. 383-404.
- Kürti, L., 2001. *The remote borderland. Transylvania in the Hungarian imagination*. Albany: State University of New York Press.
- Lefebvre, H., 2003. *The Urban Revolution*. Minneapolis: University of Minnesota Press.
- Lennon, J., Mathews, S., 1996. *Cultural landscape management. Guidelines for identifying, assessing and managing cultural landscapes in the Australian Alps national parks*. Cultural heritage working group – Australian Alps Liaison Committe.
- Lewis, P., 1979. Axioms for reading the landscapes. In: Meinig, D., W., (ed.), *The interpretation of Ordinary Landscapes*. New York: Oxford University Press, pp. 11-33.
- Longatti, P., Dalang, T., 2007. The meaning of "Landscape" - an exegesis of Swiss Government texts. In: Kienast, F. et. al., *A changing world. Challenges for landscape research*. Dordrecht: Springer, pp. 35-47.
- Longstreth, R., (ed.) 2008. *Cultural landscapes: balancing nature and heritage in preservation practice. Minneapolis*: University of Minnesota Press.
- Lorino, P., 2010. The Bakhtinian theory of chronotope (spatial-temporal frame) applied to the organizing process. Rhodes, Grecia: 2nd International Symposium on Process Organization Studies. 11-13 Iunie 2010.
- Lossi, A., 2010. Metaphor. In: Sepp, H., R., Embree, L., 2010. *Handbook of phenomenological aesthetics*. Heidelberg: Springer, pp. 211-215.
- Löwy, M., 2009. *Morning star: surrealism, marxism, anarchism, situationism, utopia*. Austin: University of Texas Press.
- Lynch, K., 1984. *Good city form*. Cambridge: MIT Press.
- Lyotard, J-F., 2003. *Conditia postmoderna*, Cluj-Napoca: Editura Idea.
- Macarie, Gh., 1980. *Geografie literară. Orizonturi spirituale in proza românească*. Bucureşti: Editura Albatros.

- Maciocco, G., 2008. *Urban Landscape Perspectives*. Springer.
- Madgearu, Al., 2005. *The Romanians in the Anonymous Gesta Hungarorum. Truth and fiction*. Cluj-Napoca: Romanian Cultural Institute.
- Marx, K., Engles, F., 1970. *The German Ideology*. International Publishers Co.
- Massumi, B., 2002. *Parables for the virtual*. Londra: Duke University Press.
- Matei, A., 2004. *Identitate culturală locală*. Cluj-Napoca: Editura U.T. PRESS.
- Mathewson, K., 1999. Cultural landscape and ecology II: regions, retrospects, revivals. In: *Progress in Human Geography*, 23 (2), pp. 267-281.
- Mărculeţ, V., Mărculeţ, I., 2011. *Mediaş. Istorie, demografie, personalităţi*. Brăila: Editura Sfântul Ierarh Nicolae.
- McCarthy, J., Wright, P., 2004. *Technology as experience*. London: The MIT Press.
- McDonough, T., 2002. Situationist Space. In: Tom McDonough, Guy *Debord and the Situationist International: texts and documents*. Cambridge: MIT Press.
- Meinig, D. W., 1979. The beholding eye. In: Meinig, D., W., ed., *The interpretation of Ordinary Landscapes*. New York: Oxford University Press, pp. 33-50.
- Meisenheimer, W., 2007. *Choreography of the Architectural Space. The disappearance of space in time*. Seul: Dongnyok Publishers.
- Mekking, A., 2009. The Architectural Representation of Reality: The Built Environment as the Materialization of a Mental Construct. In: Mekking, A. and Roose E. (ed.). *The Global Built Environment as a Representation of Realities*. Pallas Publications, pp. 23-51.
- Mercer, C., 2006. *Cultural Planning for urban development and creative cities*. Paper for Shanghai Cultural Planning Conference, 2006.
- Merleau-Ponty, M., 2009. *Phenomenology of Perception*. New York: Routledge.
- Miah, S., 2006. *Russell's theory of perception: 1905-1919*.London: Continuum International Publishing Group.
- Mihali, C., 2001. *Altfel de spaţii. Studii de heterotopologie*. Bucureşti: Editura Paideia.
- Mihali, C., 2001. *Inventarea spaţiului. Arhitecturi ale existenţei cotidiene*. Bucureşti: Editura Paideia.

- Mihali, C., Copoeru, I., 2002. *Cartografii ale modernităţii*. Cluj-Napoca: Casa Cărţii de Ştiinţă.
- Mikoleit, A., Puerckhauer, M., 2011. *Urban Code. 100 Lessons for Understanding the City*. Zurich: MIT & gta verlag.
- Mitchell, D., 2000. *Cultural geography. A critical introduction*. Oxford: Blackwell Publishers.
- Mitchell, W., 2007. Action in the hands of the user. In: Von Borries, F., et. al., 2007. *Space Time Play. Computer games, architecture and urbanism: the next level*. Boston: Birkhaeuser.
- Mitchell, W., 2008. *World's greatest architect. Making, meaning and network culture*. Cambridge: MIT Press.
- Mumford, L., 2008. Retrospect and Prospect. In: Meagher, S. M., (ed.). *Philosophy and the City*. Albany: State University of New York Press, pp. 125-133.
- Muntanola, J. T., 2004. *The child in the City: towards a dialogical model of children – environment relationships*. London: 2nd International Congress of the Child in the City Foundation, October 2004. Document published online: www.pa.upc.edu/varis/altres/arqs/05child_the_city.pdf.
- Murdoch, J., 2006. *Post-structuralist geography. A guide to relational space*. London: Sage Publications.
- Myga-Piatek, U., 2008. Between tradition and modernity of the cultural landscape research. Discussions on methodology. In: *Methodology of landscape research, Dissertations Commission of Cultural Landscape*. No. 9, Commission of Cultural Landscape of Polish Geographical Society, Sosnowiec, 2008, pp. 75-91.
- Naik, D., Oldfield, T., 2010. The Urban Industry and its Post-Critical condition. In: D. Naik, T. Oldfield. *Critical Cities. Ideas, knowledge and agitation from emerging urbanists*. London: Murdle Court Press.
- Nicolae, I., Suditu, B., 2008. *Toponimie românească şi internaţională*. Bucureşti: Editura Meronia.
- Niessen, J. P., 2005. Romania. In: R. Frucht, ed. 2005. *Eastern Europe. An introduction to the people, lands and culture, Volume 3: Southeastern Europe*. Santa Barbara: ABC-CLIO Inc., pp.735-791.
- Nieuwenhuys, C., 2002. A different City for a different Life, 1959. In: Tom McDonough, *Guy Debord and the Situationist International: texts and documents*. Cambridge: MIT Press.
- Noica, C., 2008. *Jurnal de idei*. Bucureşti: Editura Humanitas.
- Noica, C., 2010. *Pagini despre sufletul românesc*. Bucureşti: Editura

Humanitas.

- Noica, C., 2012. *Jurnal Filozofic*. Bucureşti: Editura Humanitas.
- Norberg-Schulz, C., 1979. *Genius loci. Towards a phenomenology of architecture*. New York: Rizzoli.
- O'Flaherty, M., 2010. *Cultural landscapes – is the term not redundant?*. Natural Resource Institute, University of Manitoba. [online] Accesible at the address: www.culturallandscapes.ca/essays/cultural-landscape-is-the-term-not-redundant/
- Ørstavik, I. T. B., 2005. *Forms of time and chronotope in computer programming: Run-time as adventure time?*. Conference proceedings from the XII Bakhtin Conference, Jyväskylä, Finland.
- Osmond, P., 2010. The Urban Structural Unit: towards a descriptive framework to support urban analysis and planning. In: Urban Morphology, vol. 14, no. 1.
- Paetzold, H., 2011. Rethinking Key-concepts of Modern Urban Culture: Flânerie and the City as Theater. In: Diaconu, M., et. al., 2011. *Senses and the City. An interdiciplinary approach to urban sensescapes*. Vienna: Lit Verlag, pp. 33-49.
- Palermo, P. C., 2008. Thinking over Urban Landscapes. Interpretations and Courses of Action. In: Maciocco, G. ed. *Urban Landscapes Perspective*. Springer, pp. 27-43.
- Pallasmaa, J., 2005. *The Eyes of the Skin. Architecture and the Senses*. West Sussex: John Wiley & Sons Ltd.
- Palm, R., 2009. *Hegel's concept of sublation. A critical interpretation*. Leuven: Katholieke Universiteit Leuven.
- Park, R. E., 1926. The Concept of Position in Sociology. In: Publications of the American Sociological Society, Vol. 20, pp. 1-14.
- Pirenne, H., 1946. *Medieval cities. Their origins and the revival of trade*. 4Th ed. Translated from French by Frank D. Halsey. Princeton: Princeton University Press.
- Plato, 2008. *Timaeus and Critias*. Oxford: Oxford University Press.
- Poe, E. A., 1840. *Man of the Crowd*. [online] Published online in 1994. Accessible at the address: http://etext.virginia.edu/etcbin/toccer-new2?id=PoeCrow.sgm&images=images/modeng&data=/texts/english /modeng/parsed&tag=public&part=all [accessed in 20.08.2010].
- Popescu, T., 2001. *Estetica paradoxismului*. Râmnicu Vâlcea: Editura Offsetcolor.
- Popescu, T., 2009. *Frumosul natural şi frumosul artistic*. Cluj-Napoca: Casa Cărţii de Ştiinţă.

- Pounds, N., 2005. *The Medieval City*. Westport: Greenwood Press.
- Rad, I., 2008. *Cum se scrie un text ştiinţific. Disciplinele umaniste*. Bucureşti: Polirom.
- Rath, A., 2010. Chronotopes of „Places" and „Non-Places": ecopoetics of Amitar Ghosh's The Hungry Tide. In: ASIATIC, vol. 4, no. 2.
- Rhoads, B. L., Wilson, D., 2010. Observing our World. In: Gomez, B., Jones, J. P., 2010. *Research methods in geography. A critical introduction*. Oxford: Wiley Blackwell, pp. 26-41.
- Ricoeur, P., 1985. *Time and narrative*. Volume 2. Translated from French by Kathleen McLaughlin and David Pellaver. Chicago: University of Chicago Press.
- Ringer, G., 2003. Introduction. In: Ringer, G., (ed.), 2003. *Destinations: cultural landscapes of tourism*. London: Routledge, pp. 1-15.
- Robertson, I., Richards, P., 2003. Introduction. In: Robertson, I., Richards, P., 2003. *Studying cultural landscapes*. London: Arnold Publishers. pp. 1-17.
- Robins, K., 2006. *The Challenge of Transcultural Diversities*. Strasbourg: Council of Europe Publishing.
- Robson, M., Stockwell, P., 2005. *Language in Theory. A resource book for students*. Oxon: Routledge.
- Ross, K., 2002. Lefebvre on the Situationists: an interview. In: Tom McDonough, *Guy Debord and the Situationist International: texts and documents*. Cambridge: MIT Press.
- Rubenstein, J., 2003. *Cultural landscapes. An introduction to human geography*. 7th Edition. New Jersey: Prentice Hall.
- Sadler, S., 1999. *The Situationist City*. Cambridge: MIT Press.
- Sandu, M. Al., 2007. *Eseuri de urbanism şi arhitectură. Volume 1: Sub semnul pradoxurilor = Under the sign of pradoxes*. Translated from Romanian by Magda Teodorescu and Otilia Blaga. Bucureşti: Editura Fundaţiei Arhitext design.
- Sandu, M. Al., 2007. *Eseuri de urbanism şi arhitectură. Volume 2: Sâmburele care face să crească = The Kernel that makes it grow*. Bucureşti: Editura Fundaţiei Arhitext Design.
- Sauer, C. O., 1963. The Morphology of Landscape. In: John Leighly (ed.*), Land and Life. A Selection from the Writing of Carl Ortwin Sauer*. Berkley: University of California Press, pp. 315-350.
- Sârbu, C., 2011. Peisaj cultural şi dezvoltare – substanţă şi expresie a dinamicii teritoriului. In: Sârbu, C., (ed.), 2011. *Peisaj cultural şi dezvoltare*. Bucureşti: Editura Universitară „Ion Mincu". pp. 7-15.

- Schein, R., 2004. Cultural Traditions. In: Duncan, J., et. al., 2004. *A companion to cultural geography*. Oxford: Blackwell. pp. 11-24.
- Schein, R., 2005. Syllabus. Geography 711: Cultural Landscapes. [online] Accessible at the address: www.uky.edu/AS/Geography/People/Faculty/Schein/Schein_Geo711.pdf [accessed in 7.04.2012].
- Schein, R., 2010. Cultural Landscapes. In: Gomez, B., Jones, J. P., 2010. *Research methods in geography. A critical introduction*. Oxford: Wiley Blackwell. pp. 222-241.
- Schreiber, W., Stoica, F., 2008. *Peisaje culturale istorice*. Cluj-Napoca: Argonaut.
- Scruton, R., 2011. *Cultura Modernă pe înţelesul oamenilor inteligenţi*. Translated from English by Dragoş Dodu. Bucureşti: Humanitas.
- Seager, W., 1999. *Theories of consciousness. An introduction and assessment*. London: Routledge.
- Self, W., 2007 *Psychogeography*. London: Bloomsbury.
- Semple, E. C., 2010. *Influences of Geographic Environment on the Basis of Ratzel S System of Antropo Geography*. Memphis: General Books.
- Sennett, R., 2001. A flexible City of Strangers. [online] In: *Le Monde Diplomatique* English version (February 2001). Accessible at the address: http://mondediplo.com/2001/02/16cities [accessed in 20.12.2009].
- Sharr, A., 2009. *Heidegger for Architects*. London: Routledge.
- Shaw, I. G. R., Dixon, D. P., Jones, J. P., 2010. Theorizing our world. In: Gomez, B., Jones, J. P., 2010. *Research methods in geography*. A critical introduction. Oxford: Wiley Blackwell. pp. 9-26.
- Shields, R., 1988. *Images of spaces and places. A comparative study*. University of Sussex D.Phil thesis.
- Shurmer-Smith, P., 2002. The trouble with theory. In: Shurmer-Smith, P., (ed.) 2002. *Doing cultural geography*. London: Sage Publications. pp. 11-19.
- Simmel, G., 2008. The Metropolis and Mental Life. In: Meagher, S. M., ed. *Philosophy and the City*. Albany: State University of New York Press, pp. 96-101.
- Simon, B., 2006. *Beyond Cyberspatial flâneurie*. On the Analythic Potential of Living with Digital Games, Games and Culture, volume 1 (1), pp. 62-67.
- Smethurst, P., 2000. *The Postmodern Chronotope. Reading Space and*

Time in Contemporary fiction. Amsterdam: Rodopi B.V.

- Soja, E., W., 1989. *Postmodern Geographies. The Reassertion of Space in Critical Social Theory*. London: Verso.
- Soutar, A. L., 2003. *The Chronotope as a model for Hypermedia in architecture education*. Auckland: University of Auckland Press.
- Southall, A., 1998. *The city in time and space*. Cambridge: Cambridge University Press.
- Spengler, O., 1944. *The Decline of the West. Volume 1: Form and Actuality*. Translated from German by Charles Francis Atkinson. New York: Alfred A Knopf.
- Spirn, A. W., 1998. *The language of landscape*. New Haven: Yale University Press.
- Stahl, P. S., 2005. *Case şi acareturi din Mărginimea Sibiului 1953-1958*. Bucureşti: Editura Paideia.
- Stănculescu, F., 1935. *Stil romanesc şi stil modern*. Revista Arhitectura, 1/1935, pp. 13-14.
- Stout, F., 2000. Visions of a New Reality: The City and the Emergence of Modern Visual Culture. In: Le Gates, R. T., et. al. ed. *The City Reader*. 2nd ed. New York: Routledge, pp. 143-149.
- Swarts, D., 1997. *Culture and Power. The sociology of Pierre Bourdieu*. Chicago: University of Chicago Press.
- Swyngedouw, E., 2002. *The strange respectability of the Situationist City in the Society of Spectacle*. [online] Published firstly in: International Journal of Urban and Regional Research 26(1) 2002, pp. 153-165. Accessible at the address: http://eprints.ouls.ox.ac.uk/archive/00000904/. [accessed in 14.09.2010].
- Taylor, T., L., 1999. Life in Virtual Worlds: Plural Existence, Multimodalities, and other Online Research Challenges. American Behavioral Scientist, 43 (3), Sage Publishing, pp. 436-449.
- Tănase, Al., 1968. *Introducere In filosofia culturii. Valoare Cultură Civilizaţie*. Bucureşti: Editura Ştiinţifică.
- Tihanov, G., 2000. Culture, form, life: the early Lukacs and the early Bakhtin. In: C. Brandist, G. Tihanov, ed. 2000. *Materializing Bakhtin. The Bakhtin Circle and social theory*. London: Macmillan Press Ltd. pp.43-70.
- Tilley, C., 1994. *A phenomenology of landscape. Places, paths and monuments*. Oxford: Berg Publishers.
- Trask, R. L., 2010. *Language. The basics*. Oxon: Routledge.

- Tuan, Y. F., 1977. *Space and Place. The perspective of experience*. Minneapolis: University of Minnesota Press.
- Tuan, Y. F., 2008. *Space and Place. The Perspective of Experience*. Minneapolis: University of Minnesota Press.
- UNESCO, Operational Guide 2011/WHC 11/01, Operational guidelines for the implementation of the World Heritage Convention.
- Vaneigem, R., 2002. Comments against Urbanism (1961). In: Tom McDonough, 2002. *Guy Debord and the Situationist International: texts and documents*. Cambridge: MIT Press.
- Vianu, T., 1982. *Studii de filosofia culturii*. Bucureşti: Editura Eminescu.
- Vidler, A., 2008. *Histories of the Immediate Present. Inventing Architectural Modernism*. Cambridge: MIT Press.
- Voeckler, K., 2008. Balkanology and the future of the European City. In: SAM Swiss Architecture Museum, no. 6. pp.8-11.
- Wearing, S., et. al., 2010. *Tourist Cultures: Identity, Place and the Traveler*. London: Sage Publications Ltd.
- Weber, M., 2008. Concepts and Cathegories of the City. In: Meagher, S. M., ed. *Philosophy and the City*. Albany: State University of New York Press, pp. 102-109.
- Werner, J., 2004. *American flâneur: the cosmic physiognomy of Edgar Allan Poe*. New York: Routledge.
- West-Pavlov, R., 2009. *Space in Theory. Kristeva, Foucault, Deleuze*. New York: Rodopi.
- Whyte, I. D., 2002. *Landscape and History since 1500*. London: Reaktion Books Ltd.
- William, R. D., 1998. *Aristotel.*, Bucureşti: Editura Humanitas.
- Williams, J., 2005. *Understanding Poststructuralism*, Bucks: Acumen Publishing Limited.
- Wilson, C., 2003. *Everyday America. Cultural Landscape Studies after J.B. Jackson*. Los Angeles: University of California Press.
- Zukin, S., 2000. Whose City? Whose Culture? In: Richard T. Le Gates, et al., 2000. *The City Reader*, Second Edition. New York: Routledge, pp. 131-143.

14 Notes on the Author

Alexandru Calcatinge is a young architect, urban planner and researcher on cultural landscapes. He finished his PhD in September 2012, having the main research interest in cultural landscapes theory. This book is the result of his four years doctoral studies at the University of Architecture and Urbanism "Ion Mincu" in Bucharest, Romania (with the thesis title "Contributions to the theoretical substantiation to the studies of the concept of cultural landscape"). In his second doctoral year, after the scholarship he had at the University of Vienna, he published his first book in which he presented his first incipient research on the matter of cultural landscapes as visions of the real. During the four year research, he was an academic and teaching assistant at the above mentioned University, in the Department of Urban and Landscape Design, and published several research papers on cultural landscapes, perceiving his main interest on theory.

As a short biography, it is relevant to say that he received his bachelor degree in architecture from the Technical University of Cluj-Napoca in 2007. He received his master's degree (M.Sc.) in urban planning and regional development from the University of Architecture and Urbanism "Ion Mincu" in Bucharest in 2009. He is an active member of ECOVAST Austria since 2010, a CIVILSCAPE delegate since 2011 and also a member of ISUF since 2010.

More information you can find on the following websites:

- www.visionsofthereal.com – the website of his first book
- www.peisajcultural.ro – the bilingual website dedicate by the author to his cultual landscape studies.